AF799477

Analog Circuits and Signal Processing

Series Editors

Mohammed Ismail, Khalifa University, Dublin, OH, USA

Mohamad Sawan, 18, Shilongshan Road, School of Engineering, Westlake University, Hangzhou, Zhejiang, China

The *Analog Circuits and Signal Processing* book series, formerly known as the *Kluwer International Series in Engineering and Computer Science*, is a high level academic and professional series publishing research on the design and applications of analog integrated circuits and signal processing circuits and systems. Typically per year we publish between 5-15 research monographs, professional books, handbooks, and edited volumes with worldwide distribution to engineers, researchers, educators, and libraries.The book series promotes and expedites the dissemination of new research results and tutorial views in the analog field. There is an exciting and large volume of research activity in the field worldwide. Researchers are striving to bridge the gap between classical analog work and recent advances in very large scale integration (VLSI) technologies with improved analog capabilities. Analog VLSI has been recognized as a major technology for future information processing. Analog work is showing signs of dramatic changes with emphasis on interdisciplinary research efforts combining device/circuit/technology issues. Consequently, new design concepts, strategies and design tools are being unveiled. Topics of interest include: Analog Interface Circuits and Systems; Data converters; Active-RC, switched-capacitor and continuous-time integrated filters; Mixed analog/digital VLSI; Simulation and modeling, mixed-mode simulation; Analog nonlinear and computational circuits and signal processing; Analog Artificial Neural Networks/Artificial Intelligence; Current-mode Signal Processing; Computer-Aided Design (CAD) tools; Analog Design in emerging technologies (Scalable CMOS, BiCMOS, GaAs, heterojunction and floating gate technologies, etc.); Analog Design for Test; Integrated sensors and actuators; Analog Design Automation/Knowledge-based Systems; Analog VLSI cell libraries; Analog product development; RF Front ends, Wireless communications and Microwave Circuits; Analog behavioral modeling, Analog HDL.

Tuur Van Daele • Filip Tavernier

Fully Integrated High-Voltage DC-DC and AC-DC Conversion

Springer

Tuur Van Daele
ICsense
Leuven, Belgium

Filip Tavernier
KU Leuven
Leuven, Belgium

ISSN 1872-082X ISSN 2197-1854 (electronic)
Analog Circuits and Signal Processing
ISBN 978-3-031-78518-4 ISBN 978-3-031-78519-1 (eBook)
https://doi.org/10.1007/978-3-031-78519-1

© The Editor(s) (if applicable) and The Author(s), under exclusive license to Springer Nature Switzerland AG 2025

This work is subject to copyright. All rights are solely and exclusively licensed by the Publisher, whether the whole or part of the material is concerned, specifically the rights of translation, reprinting, reuse of illustrations, recitation, broadcasting, reproduction on microfilms or in any other physical way, and transmission or information storage and retrieval, electronic adaptation, computer software, or by similar or dissimilar methodology now known or hereafter developed.
The use of general descriptive names, registered names, trademarks, service marks, etc. in this publication does not imply, even in the absence of a specific statement, that such names are exempt from the relevant protective laws and regulations and therefore free for general use.
The publisher, the authors and the editors are safe to assume that the advice and information in this book are believed to be true and accurate at the date of publication. Neither the publisher nor the authors or the editors give a warranty, expressed or implied, with respect to the material contained herein or for any errors or omissions that may have been made. The publisher remains neutral with regard to jurisdictional claims in published maps and institutional affiliations.

This Springer imprint is published by the registered company Springer Nature Switzerland AG
The registered company address is: Gewerbestrasse 11, 6330 Cham, Switzerland

If disposing of this product, please recycle the paper.

Preface

High-voltage (48 V–400 V) power sources are prevalent in our surroundings. The AC mains is found along the walls and ceilings of buildings, and DC buses are shifting to higher voltages such as 48 V and 400 V in sectors such as the automotive industry and data centers. In contrast, the growing demand for low-power applications calls for low DC supply voltages ($\leq$5 V). This trend is driven by the push for enhanced functionality and decentralization, which requires small distributed nodes equipped with sensors, actuators, and microcontrollers. Applications range from internet-of-things (IoT) nodes powered by the AC mains to control circuitry and other peripheries in electric cars supplied with high DC voltages.

To bridge the gap between high-voltage power sources (DC and AC) and low-voltage applications, a voltage converter is required. This work focuses on input voltages in the range of 400 V, which aligns with the AC mains voltage and high-voltage DC buses in electric cars and industrial environments. The intended output power is the milliwatt range, suitable for applications such as home automation, grid infrastructure, LEDs, and control circuits.

Existing solutions for 400 V with milliwatt output power are moving toward integration to improve cost-efficiency, compactness, and reliability. This evolution is characterized by an increased switching frequency to reduce the passive component size. However, the most integrated approach still requires two external components, including a cumbersome inductor. Further downsizing is constrained by the extremely low duty cycle imposed by the high conversion step, restricting higher switching frequencies. In addition, the reliance on an inductor is unsuitable as its quality deteriorates significantly when integrated.

This work explores switched-capacitor converters (SCCs) as a pathway to full integration. SCCs are exclusively composed of switches and capacitors, instead of an inductor, which makes them well-suited for integration. Furthermore, SCCs maintain a high duty cycle irrespective of the conversion step, enabling high switching frequencies to facilitate miniaturization. Despite these advantages, state-of-the-art fully integrated SCCs are limited to input voltages up to 42 V and lack compatibility with AC inputs. This book aims to overcome these two critical shortcomings in the following three parts.

The initial part elevates the maximum input voltage of fully integrated DC-DC converters from 42 V to 400 V. However, this increase brings forth challenges such as compromised component quality and heightened parasitic losses. A theoretical analysis reveals how components should block the high voltage to alleviate the reduced component quality. Furthermore, a sidewall coupling reduction technique is introduced to mitigate the parasitic losses. The resulting high-voltage topology, implemented with customized high-voltage capacitors and drivers, leads to the complete integration of a 400 V-to-12 V DC-DC converter in a 180 nm CMOS SOI technology. The prototype achieves a power density of 119 mW/mm^2 at 63.6% efficiency. This advances the input voltage of fully integrated DC-DC converters by 9$\times$ and the power density by 270$\times$, while achieving the highest efficiency in class.

The second part extends the fixed 400 V DC input voltage of the first part into a wide DC input range. This not only lays the foundation for accommodating AC input voltages in the subsequent part but also supports applications with variable DC input voltages. To mitigate the losses associated with the broad and high input voltage range, a multi-ratio topology is developed. Furthermore, this part proposes sensing and ripple reduction techniques suitable for high voltage. Measurements on a prototype showcase a 31–325 V DC input range, while achieving a 16.4 mW/mm^2 power density at 57.3% efficiency.

In the final part, the multi-ratio SCC of the preceding part is extended into an AC-DC converter, unlocking mains-powered applications. Given the substantially lower frequency of the AC mains compared to SCCs, the AC input effectively acts as DC for the SCC. Thus, after a bridge rectifier, the multi-ratio SCC can follow the AC input despite its DC-DC characteristic. Nonetheless, its minimum input voltage of 31 V requires a considerable buffer capacitance to bridge the zero-crossing of the AC wave. To address this, an AC-DC topology is introduced that reduces the size of the required buffer capacitor. This culminates in the realization of the first fully integrated AC-DC converter in the milliwatt power range. The resulting power density measures 9 mW/mm^2 at 55.1% efficiency, advancing the power density of prior fully integrated AC-DC converters by >5000$\times$.

Leuven, Belgium
August 2024

Tuur Van Daele
Filip Tavernier

Acknowledgments

The authors want to thank VLAIO for the Baekeland mandate HBC.2018.2094, Melexis for the technical and financial support, and X-FAB for the interesting discussions.

Contents

Abbreviations

AC	Alternating Current
ADC	Analog-to-Digital Converter
BCD	Bipolar-CMOS-DMOS
BOX	Buried Oxide
BP	Bottom Plate
CE	Charge Event
CMOS	Complementary Metal-Oxide-Semiconductor
CP	Charge Pump
cst	Constant
DC	Direct Current
DTI	Deep Trench Isolation
DVFS	Dynamic Voltage and Frequency Scaling
EMI	Electromagnetic Interference
ESD	Electrostatic Discharge
ESR	Equivalent Series Resistance
FET	Field Effect Transistor
FoM	Figure of Merit
FSL	Fast-Switching Limit
GaN	Gallium Nitride
HD	High-Power-Density Regime
HV	High Voltage
I/O	Input/Output
IoT	Internet of Things
iVCR	Ideal Voltage Conversion Ratio
LD	Low-Power-Density Regime
LDO	Low-Dropout Regulator
LED	Light-Emitting Diode
LS	Level-Shifter
MIM	Metal-Insulator-Metal
MOM	Metal-Oxide-Metal
MOS	Metal-Oxide-Semiconductor

MPU	Microprocessor Unit
MPW	Multi-Project Wafer
nMOS	N-channel Metal-Oxide-Semiconductor
PCB	Printed Circuit Board
PFM	Pulse Frequency Modulation
pMOS	P-channel Metal-Oxide-Semiconductor
PWM	Pulse Width Modulation
RC	Resistance-Capacitance (time constant)
RF	Radio Frequency
RMS	Root Mean Square
SAR	Successive Approximation Register
SC	Switched-Capacitor
SCC	Switched-Capacitor Converter
SiC	Silicon Carbide
SOI	Silicon-On-Insulator
SOS	Silicon-On-Sapphire
SSL	Slow Switching Limit
ST	Stage of an SCC
TP	Top Plate
VCR	Voltage Conversion Ratio

Symbols

A	Area
$a_{c,i}$	Charge multiplier element of flying capacitor i
$A_{C,k}$	Optimized area of flying capacitor k
A_{conv}	Active converter area
$a_{r,i}$	Charge multiplier element of power switch i
$A_{sw,k}$	Optimized area of power switch k
C	Capacitance
C_{boot}	Bootstrap capacitance
C_{BOX}	Buried oxide capacitance
C_{BP}	Bottom-plate substrate capacitance of a flying capacitor
C_{buf}	Buffer capacitance
C_{CP}	Charge pump capacitance
C_{DC}	DC capacitance
C_{depl}	Depletion capacitance
C_{ds}	Drain-source parasitic capacitance of a transistor
C_{fly}	Flying capacitance
C_{gs}	Gate-source parasitic capacitance of a transistor
C_j	Junction capacitance
C_{out}	Decoupling capacitance at the output
C_{par}	Parasitic capacitance of a component
C_{self}	Self-capacitance accounting for internal flying capacitors' charge time
C_{side}	Sidewall capacitance
C_{TP}	Top-plate substrate capacitance of a flying capacitor
d	Distance
D_i	Duty cycle of power switch i
$E_{dens,A}$	Area-specific energy density
E_F	Electric field
E_{loss}	Energy lost by charging the capacitor
E_{src}	Energy extracted from the voltage source
E_{sw}	Activation energy of a power switch
f_{AC}	Mains frequency

$f_{clk,CP}$	Charge pump clock frequency
$f_{clk,in}$	Input control clock frequency
$f_{clk,out}$	Output control clock frequency
f_{lin}	Linear regulator switching frequency
f_{max}	Maximum switching frequency of the SCC
f_{sw}	Switching frequency of the SCC
I_{on}	Output load current during the on-time of the linear regulator
I_{out}	Output load current
M	Number of interleaved fragments
N	Ideal conversion step
N_{max}	Maximum ideal voltage conversion step
P_{Cside}	Sidewall coupling losses
P_{DCDC}	Normalized converter losses
P_{dens}	Power density
P_{FSL}	Fast-switching limit conduction losses
P_{in}	Input power of the converter
P_{loss}	Total converter losses
P_{max}	Maximum output power
P_N	Optimized normalized losses
P_{out}	Output power of the converter
P_{par}	Parasitic switching losses
$P_{par,C}$	Parasitic switching losses related to the flying capacitor
$P_{par,sw}$	Parasitic switching losses related to the power switch
P_{Rout}	Output impedance losses
P_{SSL}	Slow-switching limit conduction losses
$P_{sw,SCC}$	Switch-parasitic losses due to the flying capacitor swings
q	Charge flowing through a capacitor
$q_{c,i}$	Charge transferred by flying capacitor i in each phase
q_{out}	Charge transferred to the output summed for all phases
$q_{r,i}$	Charge conducted by power switch i each phase
q_{src}	Charge delivered by the voltage source
R_{config}	Combined FSL impedance of all configuration switches
R_{FSL}	Fast-switching limit output impedance
R_{load}	Load resistance
R_{on}	On-resistance
R_{out}	Output impedance
R_{power}	Combined FSL impedance of all power switches
R_{shunt}	Shunt resistor
R_{SSL}	Slow-switching limit output impedance
$S_{on,rel}$	Ratio of active to total SCC stages
t	Time
t_{buf}	Buffering time
t_{on}	On-time
V	Voltage
V_0	Initial capacitor voltage

V_{bias}	Bias voltage
V_{cap}	Capacitor voltage rating
$V_{Cfly,\Phi 1}$	Bias voltage on a flying capacitor in phase 1
$V_{Cfly,\Phi 2}$	Bias voltage on a flying capacitor in phase 2
$V_{DC,i}$	Generated DC node i in the converter
V_{dd}	Supply voltage
V_{dd1}	Driver supply
V_{dd2}	Output control supply
V_{ds}	Drain-source voltage of a transistor
V_{gs}	Gate-Source voltage of a transistor
V_{in}	Input voltage of the converter
$V_{in,max}$	Maximum input voltage
$V_{in,SCC}$	SCC input voltage (in the AC-DC converter)
V_{int}	Internal node voltage of a buck converter
V_{iso}	Isolation voltage
V_{max}	Maximum voltage
$V_{min,SCC}$	Minimum input voltage of the SCC
V_{min}	Minimum input voltage of the DC-DC converter stage
V_{nbur}	Bias voltage of the n-buried layer
V_{out}	Output voltage of the converter
$V_{out,SCC}$	SCC output voltage (in the AC-DC converter)
V_{ov}	Overdrive voltage
V_{peak}	Peak AC input voltage
V_{rect}	Rectified mains input voltage
V_{ref}	Reference voltage
V_S	Source voltage of a power switch
V_{src}	Voltage of a voltage source
V_{sw}	Switch voltage rating
V_{th}	Threshold voltage
Vol_d	Dielectric volume
W_i	Width of power switch i
α	An operating mode of the AC-DC converter
α_s	Voltage dependence exponent of switches
β	An operating mode of the AC-DC converter
γ	An operating mode of the AC-DC converter
ΔE_{cap}	Difference in stored energy in the capacitor
Δt	Transition time of a load step
ΔV_{Cside}	Voltage swing on C_{side}
ΔV_{out}	Output-voltage ripple
ΔV_{par}	Voltage swing on C_{par}
ΔV_{sub}	Voltage swing on C_{BP} and C_{TP}
ϵ	Permittivity
η	(Power Conversion) Efficiency

η_{ideal}	Ideal efficiency, which is without extrinsic losses
η_{SCC}	SCC efficiency (in the AC-DC converter)
κ_{τ}	Amount of flying capacitance functioning as additional decoupling
Φ_1	Converter clock phase 1
Φ_2	Converter clock phase 2

Chapter 1
Introduction

1.1 Need for High-Voltage Conversion

High-voltage power sources surround us, yet emerging applications call for low supply voltages. This section delves into the origins of this substantial voltage gap and sets initial requirements for bridging it.

1.1.1 High-Voltage Sources Are Widely Available

High-voltage power sources are all around us (Fig. 1.1, left), with the term high voltage being defined as a voltage between 48 V and 400 V in this work.[1] The AC mains is accessible in buildings, typically located on ceilings and walls near sockets, lighting fixtures, air conditioning units, thermostats, and various other electrical devices. The AC mains delivers peak voltages up to 339 V depending on the country [92].[2] Furthermore, DC buses and batteries are shifting toward higher voltages such as 48 V and 400 V in the automotive industry, data centers, and industrial automation [18, 31, 52, 53].

The AC mains offers numerous advantages for powering indoor applications such as the Internet of Things (IoT). It is not only widely accessible within

[1] The term high voltage needs to be defined in this work as there is no uniform standard. The defined voltage range significantly exceeds the digital core voltage of CMOS technologies ($\leq$5 V), thus requiring a high-voltage technology (cf. Sect. 1.5) and considerable isolation overhead. Such requirements limit the use of conventional low-voltage techniques. A detailed discussion and further challenges can be found in Sect. 3.3.

[2] Throughout this book, both AC peak voltages and DC voltages are denoted in volts (V), while AC root mean square (RMS) voltages are specified in volts RMS (V_{RMS}).

© The Author(s), under exclusive license to Springer Nature Switzerland AG 2025
T. Van Daele, F. Tavernier, *Fully Integrated High-Voltage DC-DC and AC-DC Conversion*, Analog Circuits and Signal Processing,
https://doi.org/10.1007/978-3-031-78519-1_1

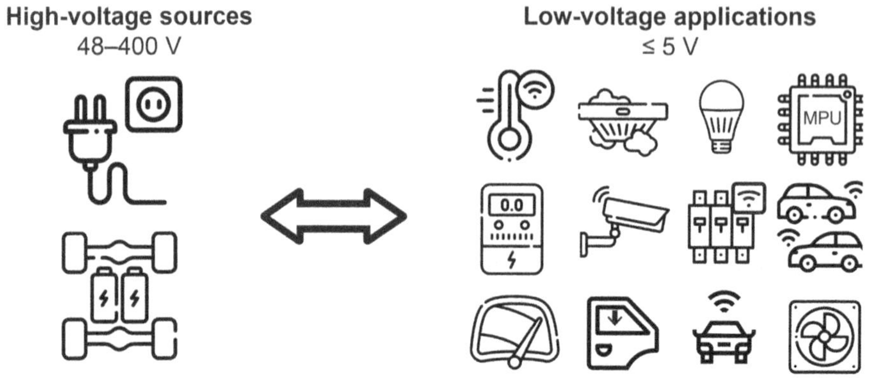

Fig. 1.1 Voltage gap between high-voltage sources and low-power applications, which require low supply voltages. (Icons made by Freepik from www.flaticon.com)

buildings but also provides a reliable and practically unlimited power source. This stands in contrast to alternative power sources like batteries and energy harvesters. Small batteries demand regular maintenance, while larger ones are both expensive and cumbersome. Energy harvesters, which extract energy from the environment (e.g., solar, radio frequency (RF), and thermal), are unreliable and offer limited power for indoor applications [74]. Their unpredictability is rooted in their reliance on the environment, which can result in sudden shutdowns without sufficient energy storage. Furthermore, ambient power sources within indoor environments are notably weaker than their outdoor counterparts. For instance, the light intensity of artificial indoor lighting is only a fraction of that provided by outdoor sunlight. This limits the harvester's power density, which typically restricts the output power to a maximum of 1 mW [59].

Regarding DC bus and battery voltages, there is a trend toward higher voltages due to the higher power demand of emerging applications. One driving force behind these rising power levels is the increasing functionality across various domains. Specifically, cars require advanced features for autonomous driving and infotainment. Similarly, servers need an increasing amount of power to enable cloud computing, artificial intelligence, and IoT. In robotics, new functionality is required for factory automation, collaborative robots, and home automation. The enhanced functionality prompts a transition from the traditional DC system voltage of 12 V to higher DC voltages such as 48 V [52] or even 400 V [31, 46, 53]. A second driving force is the electrification of mobility, propelled by emission norms and fuel economy. In hybrid and fully electric cars, the electric motor assists or replaces the traditional combustion engine. Depending on the vehicle type, this requires battery voltages reaching up to 400 V DC and beyond [18].

Adopting higher DC bus and battery voltages is critical to facilitate these increasing power demands. As power increases, so do the distribution losses, which scale quadratically with the current flowing through the distribution network. Elevating the system bus voltage reduces this current, thereby diminishing the associated losses and cable cross sections. Note that this approach is analogous to the use of high voltages in long-distance electrical power distribution.

1.1.2 Low Supply Voltages Are Required

Despite the availability of high-voltage sources (DC and AC), there is a growing demand for low-power applications (Fig. 1.1, right), which require low DC supply voltages of 5 V and below.[3] This work targets these low-power applications, which are becoming increasingly prevalent because of the trend toward enhanced functionality and decentralization. This shift necessitates numerous small distributed nodes, each equipped with sensors, actuators, microcontrollers, and/or transceivers. Given their high level of integration, these circuits need low supply voltages.

The growing low-power applications powered by the AC mains are found in IoT nodes deployed in smart homes, offices, and industrial environments. This surge requires many small mains-power nodes with embedded intelligence to monitor and control the physical environment. Within these nodes, sensors measure motion (for security), temperature, humidity, gas (for safety, e.g., smoke detectors), and more. Actuators control wireless light switches, audio, door locks, etc. Microcontrollers perform local computations required for e-metering in grid infrastructure, which accurately measures the power consumption, etc. Transceivers facilitate wireless communication between the IoT node and other devices or the central control system.

Similarly, the growth of low-power applications extends to fields like the automotive industry that use high-voltage DC buses and batteries. A multitude of small and distributed sensors helps to drive and park, incorporating features like rear collision warning, lane assistance, and pedestrian detection, with many more expected to pave the way for fully autonomous driving. Numerous voltage, temperature, and current measurements in battery management systems optimize energy usage and extend battery life. Actuators such as (pre-)drivers control window lifts, fans, wipers, vent flaps, light-emitting diodes (LEDs), etc. Transceivers enable connectivity to other vehicles and the Internet, accessing weather conditions, traffic lights, etc.

1.1.3 Bridging the Voltage Gap

Bridging the voltage gap between high-voltage sources (DC or AC) and low-voltage applications requires a voltage converter. This work addresses input voltages of

---

[3] This work defines low voltages as voltages equal to or lower than 5 V, aligning with the digital core voltage in CMOS technology.

around 400 V (DC or AC peak), corresponding to power sources such as the AC mains and high-voltage DC car batteries. However, the presented topologies are suitable for applications with 48 V DC buses and batteries as well. The targeted output power is the milliwatt range to supply applications such as home automation, grid infrastructure, LEDs, and control circuits [45, 76].

1.2 Motivation for Complete Integration

Besides the ~400 V input voltage (DC or AC peak) and milliwatt output power specification, a fully integrated solution is preferred. This brings forth numerous advantages, which can be categorized into three main groups: cost, size, and performance [8].

As the substantial number of distributed electronic circuits continues to grow, there is an increasing demand for a cost-effective solution. This demand is met by fully integrating the circuits, including the bulky voltage converter. The elimination of external components not only reduces the bill of materials but also results in a smaller printed circuit board (PCB) size, reducing the manufacturing and assembly costs. Note that prior works [11, 65] have co-packaged external components to save PCB area. However, this solution results in high packaging costs. On the other hand, the additional cost to integrate the passive components on-chip is minimal. In particular, integration allows for higher switching frequencies, making these passive components smaller while delivering the same output power. Additionally, the mass production of microchips, particularly those using silicon CMOS processes, further decreases the cost.

Another advantage of integration lies in the reduction of size, as eliminating external components reduces the PCB area and, consequently, the total system size. This is essential to fit systems within the same enclosure size despite their growing functionality [54, 76]. For example, newer-generation circuit breakers implement features such as wireless connectivity and power flow monitoring. Incorporating these additional circuits into the current chassis demands a compact solution.

Finally, integration enhances performance in various specifications and at the system level. One advantage is the ability to integrate the converter and the load onto the same chip, resulting in a fully integrated feedback loop. The reduced physical dimension of the feedback loop results in lower parasitics, enabling a higher switching frequency. This reduces the latency between successive power transfers, resulting in a faster transient response. Additionally, the integration of the DC-DC converter also allows for an increase in chip input voltage while maintaining the same power. This reduces the intake current leading to lower losses in the power delivery network. Next, integration enables granularization, where the supply voltage of each load can be optimized independently using a separate DC-DC converter. This facilitates efficiency improvement techniques such as dynamic voltage and frequency scaling (DVFS).

Nevertheless, it is important to recognize that the integration of passive components also adversely impacts the performance, given that their quality is inferior to external counterparts. For instance, integrated components suffer from parasitic capacitive coupling and can store less energy due to size constraints.

1.3 Existing Solutions

Conventional voltage converters meet the specifications of ~400 V input voltage (DC and AC peak) and milliwatt output power; however, they require multiple external components. This section centers on their trend toward integration and explains why these approaches cannot achieve the ultimate goal of complete integration.

1.3.1 Trend Toward Integration

Figure 1.2 depicts conventional solutions (and this work) bridging the voltage gap while delivering milliwatts of power. All the converters are compatible with the AC mains voltage (i.e., European 230 V_{RMS}), although the three right ones also accommodate high DC input voltages (~400 V). Progressing from left to right shows the evolution toward smaller and more integrated converters. Initially, the linear power supply on the left [29] relies on a substantial transformer. Subsequently, in the flyback converter [57], this transformer is replaced with a more compact coupled inductor (yellow component at the top-left). In the latest development [60], this coupled inductor is further replaced by a smaller inductor.

The size reduction is facilitated by operating the converter at a higher switching frequency (f_{sw}). The converter's dimensions are predominantly influenced by its passive components, which generally store and release energy during each switching

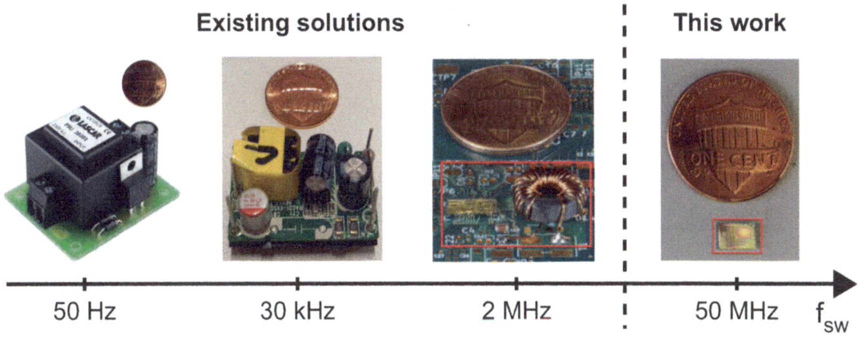

Fig. 1.2 Conventional mains-supplied AC-DC converters in the milliwatt power range [29, 57, 60] and this work

cycle. Elevating the switching frequency reduces the amount of energy that needs to be stored per cycle, resulting in smaller passive components. This is visually demonstrated by Fig. 1.2, where the smaller converters operate at higher switching frequencies.

1.3.2 Problems Preventing Complete Integration

Despite the trend toward integration, the existing solutions cannot reach the ultimate goal of complete integration. The most integrated solution [60] still needs two external components. To comprehend its limitations, the fundamental operation of an inductive buck converter is explained, as illustrated in Fig. 1.3. Two switches generate a pulse-width-modulated (PWM) signal at the internal node (V_{int}). This PWM signal is filtered by an inductive low-pass filter, resulting in a DC output voltage (V_{out}). As the value of the output voltage corresponds to the average of the PWM signal, it is set by the duty cycle for a given DC input voltage (V_{in}).

The first obstacle to achieving complete integration is the very low duty cycle needed to bridge the large specified voltage gap. This low duty cycle impedes the increase of the switching frequency, as it would lead to an extremely low on-time. To illustrate, switching the inductive buck converter at 50 MHz while converting 400 V DC to 12 V DC would result in an on-time of only 600 ps. Beyond the control challenges, this mandates the voltage on node V_{int} to rise from 0 V to 400 V and back within 600 ps, demanding a slew rate of >1300 V/ns. In comparison, integrated high-voltage power switches in silicon technologies are confined to slew rates of about 50 V/ns [90]. This technology limitation persists irrespective of the switch size, as the lower on-resistance of larger switches is counteracted by larger parasitic capacitances.

Moreover, even if a switching frequency of 50 MHz were feasible, integrating the inductor at this frequency is unsuitable. This integration would result in a low quality factor, leading to significant performance deterioration. Additionally, the proximity of the integrated inductor to the lossy substrate further diminishes its quality.

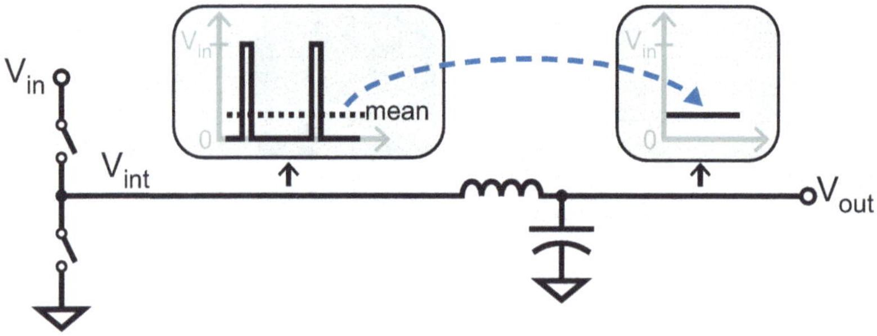

Fig. 1.3 Basic operation of an inductive buck converter

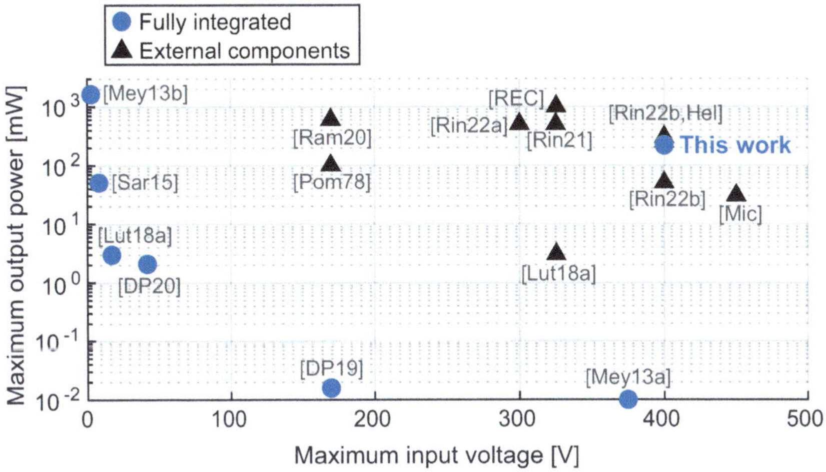

Fig. 1.4 Maximum output power versus maximum input voltage (DC or AC peak) for state-of-the-art DC-DC and AC-DC converters

Unfortunately, the quality issue is fundamental for small magnetic components [72], thereby challenging the prospects for the development of high-quality integrated magnetic components in the future.

Expanding on the discussion of existing solutions, Fig. 1.4 provides an overview of state-of-the-art publications and commercial products, encompassing solutions with both AC and DC input voltages. It shows their maximum output power versus maximum input voltage (DC or AC peak), highlighting whether they are fully integrated. Previous solutions achieving the target input voltages (~400 V DC or AC peak) and output powers (milliwatt range) invariably rely on external components, as discussed earlier. While previous fully integrated solutions achieve either a high input voltage [41] or a high output power [42, 64], none succeed in accomplishing both simultaneously. In particular, the high-power solutions are confined to an input voltage of 8 V, while the high-voltage solution is limited to a 10 μW output power and exclusively accommodates AC input voltages.

1.4 Switched-Capacitor as the Solution

Although complete integration offers numerous advantages, the integration of conventional approaches has been constrained by low duty cycles and the inductor. This work explores the switched-capacitor converter (SCC) as a solution, circumventing both issues. Specifically, the SCC relies solely on switches and capacitors, eliminating the need for an inductor. Furthermore, switched-capacitor topologies set the conversion step independent of the duty cycle, enabling high switching frequencies without necessitating extremely low on-times.

Although the SCC is a promising candidate, state-of-the-art fully integrated SCCs do not meet the required specifications. First, their input voltages are limited to 42 V [15], while target applications require input voltages of ~400 V. Second, as the SCC is a DC-DC converter, it lacks compatibility with AC inputs, which is necessary for mains-supplied applications. The primary objective of this book is to resolve these two crucial limitations.

Furthermore, the SCC should meet numerous other specifications to be viable in practical applications. While different applications prioritize other specifications, efficiency and power density are universally important [8]. Efficiency is essential to reduce energy consumption and prevent overheating, while power density directly correlates with the converter area and, consequently, its cost. A detailed discussion of the specifications can be found in Sect. 3.1.

1.5 Technology

To fully integrate a high-voltage SCC, the technology should satisfy two key criteria. First, it must be capable of handling voltages as high as 400 V (DC or AC peak). Second, the technology should provide high-speed and high-density digital circuitry. In particular, a fast digital controller enables high switching frequencies, which in turn miniaturizes the passive components (cf. Sect. 1.3.1). Furthermore, high-density digital logic is required for integrating the increasing system functionality onto the same die.

Baseline nanometer-CMOS technologies are tailored for digital logic, facilitating integrated high-frequency controllers and system functionality. Yet, they fall short of supporting the 400 V input voltage (DC or AC peak). In particular, advanced CMOS technologies like 16 nm, 7 nm, and 3 nm operate with core supply voltages below 1 V. Although techniques like stacking transistors can expand the voltage capability, their effectiveness is bounded by the well-to-substrate isolation of $\sim 10\,V$.[4]

Alternatively, wide-bandgap technologies such as gallium nitride (GaN) offer high-voltage transistors with superior performance. Nonetheless, integrating digital circuitry and drivers on the same substrate as the converter is challenging due to the lack of suitable p-type devices. A workaround uses pull-up resistors to substitute for the absent p-type device, but this results in static power consumption. To keep this consumption within acceptable limits, the integrated GaN controller operates at a reduced switching frequency, such as 1 MHz in [88] and [77].

Finally, high-voltage CMOS technologies incorporate high-voltage devices and isolation in silicon CMOS technologies, while preserving the low-voltage CMOS

[4] Note that custom high-voltage capacitors can block higher voltages in standard CMOS technologies [14, 41]. Nonetheless, their usage is confined to AC-DC converters, particularly those topologies where active devices do not interface the high voltage. Moreover, these AC-DC converters are constrained to output powers of $\leq 15\,\mu W$ (cf. Fig. 1.4), not meeting the required milliwatt output power.

transistors for digital and analog functionality. This technology is often augmented with bipolar transistors, culminating in a bipolar-CMOS-DMOS (BCD) technology. High-voltage CMOS technologies exhibit feature sizes larger than 100 nm, in contrast to the few nanometers in baseline nanometer-CMOS. This distinction originates from the core objective of high-voltage CMOS, which prioritizes the integration of additional functionality over the relentless pursuit of lithographic scaling [47], a concept known as *More than Moore*. Nonetheless, there persists a discernible trend toward technology nodes below 100 nm, motivated by the growing demand for faster and denser digital circuits and interfaces of the system [61].

Among the considered technologies, only the high-voltage CMOS one combines high-voltage handling with high-performance digital circuitry. Therefore, this work focuses on integrating high-voltage SCCs within such a technology. Notably, existing commercial technologies offer devices and isolation capabilities of 400 V and beyond within a 180 nm high-voltage CMOS process [50, 93].

1.6 Book Overview

This chapter explores the feasibility and the methodology of fully integrating DC-DC and AC-DC converters with high input voltages ($\sim$400 V) in the milliwatt power range. Chapter 1 introduced the research question and proposed the switched-capacitor converter as a promising candidate for realizing these objectives.

Before exploring the use of SCCs for high input voltages, Chap. 2 elucidates the fundamental principles of fully integrated SCCs. It covers their loss mechanisms and control strategies, which are essential for comprehending the subsequent chapters.

Despite the potential of fully integrated SCCs to fulfill the intended goals, existing works are limited to input voltages of 42 V. Chapter 3 proposes a topology and implementation to push this DC input voltage to 400 V, thereby unlocking new applications. This converter is fully integrated into a high-voltage CMOS technology, and its measurement results are provided.

Building on the foundation of the preceding chapter, Chap. 4 broadens the fixed DC input voltage to a wide DC input range. This not only enables applications demanding a broad DC input range, but also marks a first step toward compatibility with the ubiquitous AC mains.

While the converter from Chap. 4 can handle a broad range of AC input voltages, it necessitates a sizable buffer capacitor to bridge the AC zero-crossing. Chapter 5 explores techniques to minimize the size of the buffer capacitor within the SCC topology. The AC-DC converter is fully integrated into a high-voltage CMOS technology and its measurement results are presented.

Finally, Chap. 6 summarizes the presented work.

Chapter 2
Fundamentals of Fully Integrated Switched-Capacitor Converters

The preceding chapter proposed the switched-capacitor converter (SCC) as a solution for fully integrating a high-voltage converter in the milliwatt power range. Before exploring SCCs at 400 V DC input voltage, this chapter delves into the fundamentals of fully integrated switched-capacitor DC-DC converters.

Initially, it elucidates the operational principle underlying an SCC. This is followed by an analysis encompassing the equivalent model of the SCC and its losses, while also examining the trade-off in relation to power density. Concluding the chapter, control strategies are examined to manage load and line variations.

2.1 Operational Principle

The operational principle of a switched-capacitor DC-DC converter in steady state is exemplified using a basic down-converter, as illustrated in Fig. 2.1. The converter alters the provided DC input voltage (V_{in}) into a DC output voltage (V_{out}). It employs two nonoverlapping clock signals (Φ_1, Φ_2) to control four power switches. Specifically, power switches labeled Φ_1 are closed in clock phase 1, while those marked Φ_2 are on in clock phase 2. As a result, the flying capacitor (C_{fly}) alternates between two configurations, or phases, which are illustrated in Fig. 2.1b and c. In phase 1, the input charges C_{fly} with charge q, while this charge continues to flow to the output. During phase 2, the flying capacitor is connected in parallel with the output load, discharging the stored charge q to the output. The continual alternation between the two phases transfers charge from the input to the output, by means of the flying capacitor.

A sequence of certain phases constitutes a topology, establishing a fixed ideal input-output voltage ratio in steady state. This ideal ratio's value can be easily deduced from Fig. 2.1 for the basic converter. The term ideal denotes no load and all ideal components, leading to no charge flow in steady-state operation. This

© The Author(s), under exclusive license to Springer Nature Switzerland AG 2025
T. Van Daele, F. Tavernier, *Fully Integrated High-Voltage DC-DC and AC-DC Conversion*, Analog Circuits and Signal Processing,
https://doi.org/10.1007/978-3-031-78519-1_2

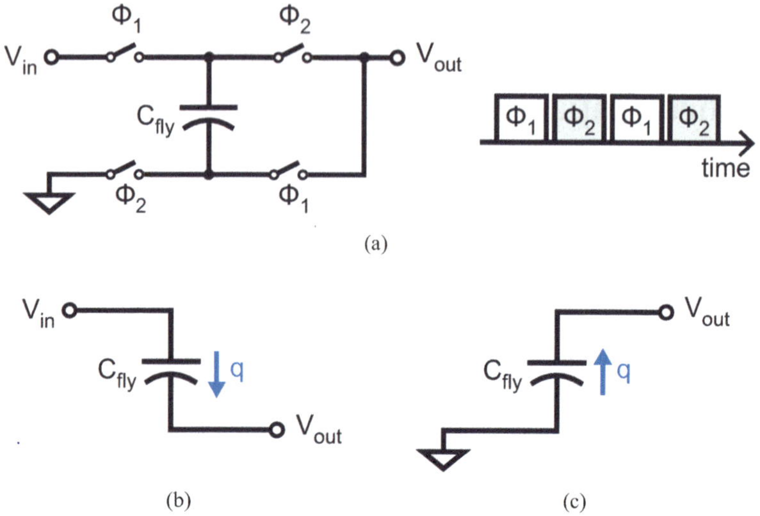

Fig. 2.1 A basic 2:1 switched-capacitor DC-DC converter. (**a**) Topology, along with clock phases versus time. (**b**) Equivalent circuit in phase 1. (**c**) Equivalent circuit in phase 2

condition maintains a constant voltage across the flying capacitors throughout the phases. Therefore, the voltages across the flying capacitor in both phases ($V_{Cfly,\Phi 1}$, $V_{Cfly,\Phi 2}$) are equal. They can be expressed by the following equations:

$$V_{Cfly,\Phi 1} = V_{in} - V_{out}, \tag{2.1}$$

$$V_{Cfly,\Phi 2} = V_{out}. \tag{2.2}$$

Equating these expressions and solving them for the ratio of V_{out} to V_{in} yield

$$iVCR = \left(\frac{V_{out}}{V_{in}}\right)_{ideal}, \tag{2.3}$$

$$= \frac{1}{2}. \tag{2.4}$$

This ratio is called the ideal voltage conversion ratio (iVCR), which is set by the topology. The iVCR of 1/2, often denoted as 2:1, indicates that the converter halves the input voltage toward the output, assuming no load and ideal components.

It is worth noting that upcoming chapters sometimes use the ideal conversion step (N) instead of the iVCR. For step-down converters, which is the focus of this work, the relationship between N and iVCR is given by

$$N = \frac{1}{iVCR}. \tag{2.5}$$

This terminology avoids confusion in describing a significant voltage drop from input to output by the converter. Denoting this as a *large conversion step* is mathematically correct, which stands in contrast to earlier works that termed it a *large conversion ratio*.

2.2 Analysis

While the operational principle was explained under idealized conditions for simplicity, this is not representative for practical implementations. Therefore, this section analyzes the equivalent circuit and the key loss mechanisms when considering nonideal components and a load. Furthermore, it investigates how power density affects these losses, given its significance.

2.2.1 *Equivalent Model*

Figure 2.2 depicts the equivalent model of an SCC in steady state [37]. The ideal DC transformer with a ratio of 1:iVCR models the ideal SCC without load, as discussed before. However, the introduction of a load (R_{load}) initiates charge transfer, involving the charging and discharging of flying capacitors, which requires a voltage drop across them. This voltage drop is proportional to the output current toward the load (I_{out}) and can thus be represented by a resistive output impedance (R_{out}).

The origins and formulas of this output impedance will be elaborated on in Sect. 2.2.2, with the current subsection discussing its implications at the circuit level.

Beyond inherent losses, the output impedance introduces an additional voltage drop when a nonzero output current is present. The reduced DC output voltage is expressed by

$$V_{out} = V_{in} \cdot iVCR - I_{out} \cdot R_{out}. \tag{2.6}$$

Fig. 2.2 Equivalent model of an SCC

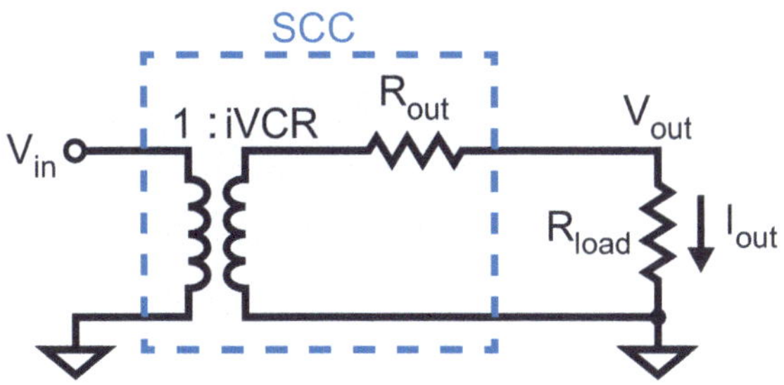

This equation underscores that the iVCR deviates from the actual voltage conversion ratio (VCR) in the case of a nonzero output current. The VCR is defined as

$$VCR = \frac{V_{out}}{V_{in}}. \tag{2.7}$$

2.2.2 *Intrinsic Losses*

The output impedance introduces intrinsic losses, with the term indicating that these losses are embedded in the equivalent model. The behavior of the output impedance depends on whether the charge transfer of the flying capacitors can reach equilibrium before transitioning to the subsequent phase. This is contingent on the time allocated for the transfer, which is inversely proportional to the switching frequency (f_{sw}) of the converter. Consequently, two distinct asymptotic behaviors arise: the slow-switching limit (SSL) and the fast-switching limit (FSL), corresponding to slow- and fast-switching frequencies, respectively.

The designations *fast* and *slow* pertain to the time constant required for charging the flying capacitors and thus reaching equilibrium. This time constant originates from the RC circuit generally formed by the flying capacitor and the on-resistance of the power switches through which it is charged. When the phase duration ($\propto 1/f_{sw}$) substantially exceeds this RC constant, the converter operates in the SSL region. Conversely, when the phase duration is considerably shorter than the RC constant, it functions in the FSL region.

2.2.2.1 Slow-Switching Limit

In the SSL region, the charge transfer to and from flying capacitors reaches equilibrium. To determine the associated losses, consider a DC voltage source charging an ideal capacitor (C) in Fig. 2.3.[1] Figure 2.3a represents the initial situation: The capacitor has an initial DC voltage V_0 and is separated by an open switch from a voltage source with voltage V_{src}. Subsequently, when the switch is closed, the capacitor is charged until the DC source voltage V_{src} is reached across the capacitor at equilibrium (Fig. 2.3b). The charge delivered by the voltage source (q_{src}) is given by

$$q_{src} = C \cdot (V_{src} - V_0). \tag{2.8}$$

[1] For simplicity, this analysis focuses on charging a single capacitor with a voltage source. However, the derived conclusions can be extended to charge transfer between multiple capacitors and/or voltage sources.

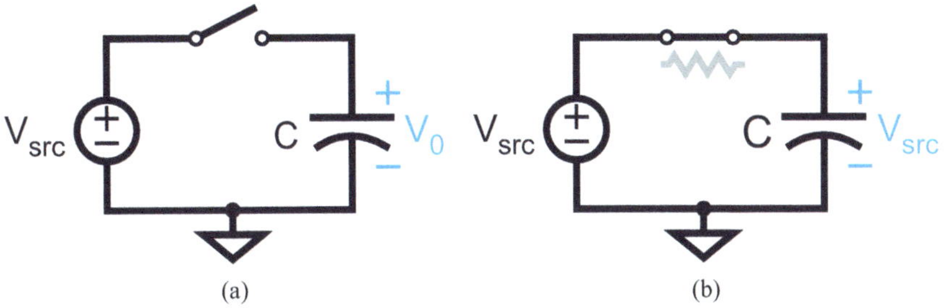

Fig. 2.3 Charging a capacitor with a voltage source. (**a**) Initial state. (**b**) Equilibrium

Using this equation, the energy extracted from the voltage source (E_{src}) is obtained:

$$E_{src} = V_{src} \cdot q_{src}, \tag{2.9}$$

$$= C \cdot V_{src} \cdot (V_{src} - V_0). \tag{2.10}$$

Furthermore, the energy added to the capacitor (ΔE_{cap}) during the charging process can be described as

$$\Delta E_{cap} = \frac{C \cdot (V_{src}^2 - V_0^2)}{2}. \tag{2.11}$$

The difference between E_{src} and ΔE_{cap} reveals the energy lost (E_{loss}) by charging the capacitor:

$$E_{loss} = E_{src} - \Delta E_{cap}, \tag{2.12}$$

$$= \frac{C \cdot (V_{src} - V_0)^2}{2}. \tag{2.13}$$

Remarkably, this energy loss remains unaffected by the on-resistance of the interconnecting switch. This is because a lower on-resistance leads to both a shorter charging time and a higher current value, counterbalancing each other. Known as charge-sharing losses, these are inherent in SCCs due to the charging and discharging of flying capacitors using voltage sources or other flying capacitors.

The charge-sharing losses in a general SCC are modeled by the SSL output impedance [66]. This impedance can be expressed exclusively in terms of design and topology variables, eliminating the direct dependence on the voltage difference $V_{src} - V_0$. For two-phase converters, it is given by:[2]

[2] Equation (2.14) assumes an infinite output capacitance, but it can be modified to incorporate the effect of a finite output capacitance [81].

$$R_{SSL} = \sum_{i \in caps} \frac{(a_{c,i})^2}{C_{fly,i} \cdot f_{sw}}, \tag{2.14}$$

$$a_{c,i} = \frac{q_{c,i}}{q_{out}}. \tag{2.15}$$

Here, $C_{fly,i}$ denotes the capacitance of the ith flying capacitor, $a_{c,i}$ represents the charge multiplier element of the ith flying capacitor given by (2.15), and $q_{c,i}$ signifies the corresponding charge transferred in each period. The charge multiplier elements express the amount of charge flowing through the flying capacitors of the SCC in each phase to deliver a certain output charge q_{out}. The elements follow from the topology and can be derived through the charge flow analysis of [66].

By substituting R_{out} with R_{SSL} in the equivalent model of Fig. 2.2, the intrinsic losses in the SSL region (P_{SSL}) are given by

$$P_{SSL} = I_{out}^2 \cdot R_{SSL}. \tag{2.16}$$

2.2.2.2 Fast-Switching Limit

Unlike the SSL region, the charge transfer to and from flying capacitors cannot achieve equilibrium in the FSL region. In the asymptotic limit, flying capacitor voltages remain constant, and their behavior can be modeled by voltage sources. Consequently, there is a constant voltage drop across the interconnecting power switches during charge transfer, causing a constant current dictated by the on-resistance. This results in ohmic losses depending on the on-resistance, in contrast to the charge-sharing losses in the SSL region. The corresponding fast-switching output impedance (R_{FSL}) for two-phase converters is given by (2.17) [66].

$$R_{FSL} = \sum_{i \in switches} \frac{R_{on,i} \cdot (a_{r,i})^2}{D_i}. \tag{2.17}$$

$$a_{r,i} = \frac{q_{r,i}}{q_{out}}. \tag{2.18}$$

Here, $R_{on,i}$ represents the on-resistance of power switch i. The symbol D_i signifies its duty cycle, which is typically 50% for two-phase converters. The term $a_{r,i}$ denotes the charge multiplier element of the ith power switch given by (2.18), with $q_{r,i}$ the corresponding charge flowing each period. The charge multiplier elements are similar to $a_{c,i}$ for flying capacitors and are, once again, determined by the converter topology. As opposed to R_{SSL}, R_{FSL} is independent of the switching frequency. Its reduction for a given topology can only be accomplished by decreasing the on-resistance of the power switches, generally by increasing their width, when assuming a constant duty cycle.

By substituting R_{out} with R_{FSL} in the equivalent model of Fig. 2.2, the intrinsic losses in the FSL region (P_{FSL}) can be determined:

$$P_{FSL} = I_{out}^2 \cdot R_{FSL}. \tag{2.19}$$

For the sake of completeness, the equivalent series resistance (ESR) of the flying capacitors and the parasitic resistance of metal interconnections also influence the value of R_{FSL}. These contributions can be added to (2.17), as demonstrated in [42]. Nevertheless, a proper layout can generally mitigate these influences, rendering them negligible. Consequently, they are not considered in this work.

2.2.2.3 Combined Output Impedance

Although R_{SSL} and R_{FSL} accurately describe the output impedance at their respective limits, their combined influence shapes the overall output impedance in the intermediate region. This behavior is commonly approximated using the Euclidean norm of the SSL and FSL impedance [66]:

$$R_{out} \approx \sqrt{R_{SSL}^2 + R_{FSL}^2}. \tag{2.20}$$

The corresponding output impedance losses (P_{Rout}) are given by

$$P_{Rout} = I_{out}^2 \cdot R_{out}. \tag{2.21}$$

Figure 2.4 plots the output impedance versus the switching frequency on a logarithmic scale, consistent with the previously discussed behavior. The graph shows the two asymptotic limits, SSL and FSL, with the intermediate region a combination of both. In the SSL region, the output impedance is inversely proportional to the switching frequency, while in the FSL region, it remains constant.

2.2.3 *Extrinsic Losses*

In addition to intrinsic losses, an SCC also incurs extrinsic losses, which are not included in the equivalent model of Fig. 2.2. These extrinsic losses primarily arise from the integration of flying capacitors and power switches in a practical implementation, giving rise to parasitic capacitors. The periodic charging and discharging of these parasitic capacitors (C_{par}) with their voltage swing (ΔV_{par}) generate parasitic switching losses (P_{par}), given by

$$P_{par} = f_{sw} \cdot C_{par} \cdot \Delta V_{par}^2. \tag{2.22}$$

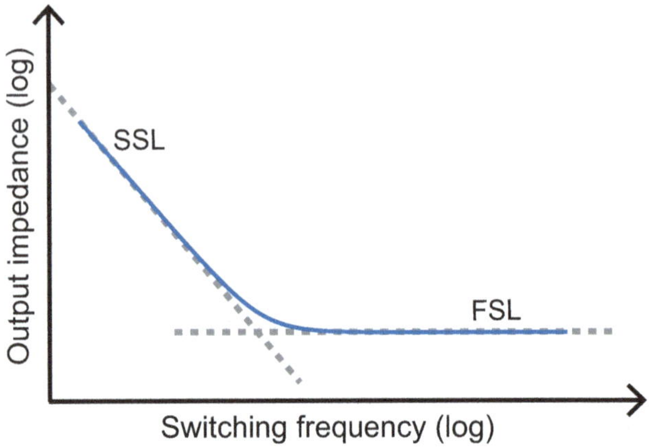

Fig. 2.4 Output impedance versus switching frequency, with dotted lines indicating the asymptotic limits

These parasitic losses are further analyzed for power switches and flying capacitors in the following subsections.

2.2.3.1 Power Switches

Implementing the power switches as transistors introduces parasitic capacitors, with predominant losses attributed to the gate-source capacitor (C_{gs}) and the drain-source capacitor (C_{ds}). The corresponding gate-source voltage swing (ΔV_{gs}) and drain-source voltage swing (ΔV_{ds}) during SCC operation result in the switch-parasitic losses ($P_{par,sw}$):

$$P_{par,sw} = f_{sw} \sum_{i \in switch} (C_{gs,i} \cdot \Delta V_{gs,i}^2 + C_{ds,i} \cdot \Delta V_{ds,i}^2), \quad \text{with } C_{gs,i}, C_{ds,i} \propto W_i, \tag{2.23}$$

where W_i is the width of the ith power switch. For example, gate-source capacitors are periodically charged to activate the power switches. Since the actual gate-source and drain-source capacitances vary with voltage, the capacitances in (2.23) represent their linear equivalents, which exhibit identical losses for the corresponding voltage swings.

Although the drain-source capacitance is often neglected in low-voltage SCCs [8, 30, 64], it becomes significant in high-voltage converters. Despite typically having a lower capacitance than the gate-source capacitance, the drain-source swing can extend to significantly higher voltages. In particular, gate-source swings are typically limited to 5 V due to the gate oxide, while drain-source voltages can reach up to 400 V in this work depending on the topology. These higher voltage swings

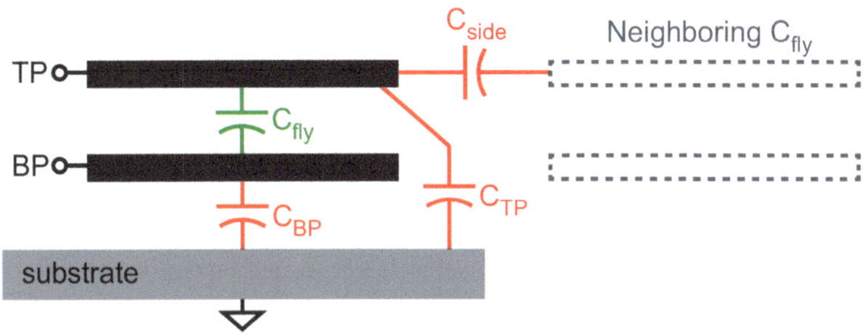

Fig. 2.5 Parasitic capacitors of an integrated sandwich capacitor

notably impact the losses due to their quadratic relationship (2.23), underscoring the imperative to include the drain-source capacitance in high-voltage SCC analysis.

2.2.3.2 Flying Capacitors

Similar to power switches, integrating flying capacitors introduces parasitic capacitors, which in this context couple to the substrate and surrounding circuits. This is illustrated in Fig. 2.5, which shows the implementation of a flying capacitor as a sandwich capacitor with two horizontal parallel plates. Despite providing useful capacitance (C_{fly}), its proximity to the grounded substrate in integrated technologies leads to top-plate capacitance (C_{TP}) and bottom-plate capacitance (C_{BP}) to this substrate.[3] Note that the terms top plate (TP) and bottom plate (BP) are defined based on the voltage levels within the SCC rather than their proximity to the substrate, with the top plate linked to the highest voltage. This distinction is important because many SCC designs position their top plate the closest to the substrate for improved efficiency and/or power density [42].

While previous works solely considered substrate capacitors [8, 30, 66], this work incorporates sidewall capacitors as well. This additional parasitic capacitor originates from the lateral coupling among adjacent flying capacitors. This becomes significant at high input voltages, where many flying capacitors are integrated close together, which is further discussed in Sect. 3.4.2. Furthermore, with blocking voltages exceeding 5 V in this work, metal-oxide-metal (MOM) capacitors (such as the sandwich type) are generally employed.[4] These MOM capacitors use all

[3] While C_{TP} is predominantly shielded by C_{BP} in the sandwich capacitor of Fig 2.5, this shielding does not universally apply to all capacitor types.

[4] MOM capacitors are also used at voltages of 5 V and below by stacking them on top of MOS capacitors to augment the capacitance density [10, 32]. This stacking approach is particularly beneficial for the higher voltages corresponding to thick-oxide transistors, where it significantly boosts the overall capacitance density.

available metal layers, resulting in a substantial sidewall area, which further augments the sidewall capacitance.

The parasitic capacitors of the flying capacitors are periodically charged during SCC operation. For example, in Fig. 2.1a, the bottom plate of C_{fly} gets charged to V_{out} in phase 1 and discharged to the ground in phase 2. This charging happens with a voltage swing modeled as ΔV_{sub} for both the top and bottom plates, as they experience approximately the same swing. Together with the voltage swing ΔV_{Cside} across the sidewall capacitors (C_{side}), this leads to the capacitor-parasitic losses ($P_{par,C}$), as formulated by

$$P_{par,C} = f_{sw} \sum_{i \in caps} \left[\left((C_{TP,i} + C_{BP,i}) \cdot \Delta V_{sub,i}^2\right) + \left(C_{side,i} \cdot \Delta V_{Cside,i}^2\right)\right]. \tag{2.24}$$

2.2.3.3 Additional Losses

Besides the parasitic losses of the power switches and flying capacitors, there is an additional extrinsic loss associated with the controller. The controller determines the activation timings of the power switches during SCC operation. Given that the implementation and losses of the controller are independent of the SCC topology, it is unnecessary to include them in the topology analysis of this work. Note that the controller's power loss is fixed, rendering it significant for efficiency under light loads, while its impact diminishes toward higher power levels.

Furthermore, the gate driver introduces yet another source of extrinsic losses. Upon receiving a control signal, the gate driver is responsible for activating (deactivating) a power switch by charging (discharging) the gate-source capacitor. The gate driver losses depend on the value of the switching frequency, the gate-source capacitance, and the gate-source voltage swing in the same way as for the switch-parasitic losses. As a result, these driver losses can be modeled by multiplying the first term in (2.23) by a factor greater than one [8].

2.2.4 *Efficiency Versus Power Density*

The four discussed loss mechanisms determine the efficiency of the converter. Acknowledging the universal importance of efficiency and power density across applications [8], this section provides their definition and investigates their interdependence.

2.2.4.1 Definitions

In the preceding subsections, four loss mechanisms of fully integrated SCCs were identified. The total converter losses (P_{loss}) are computed by aggregating these individual losses as follows:

$$P_{loss} = \sqrt{P_{SSL}^2 + P_{FSL}^2} + P_{par,sw} + P_{par,C}. \tag{2.25}$$

These total losses impact the power conversion efficiency, referred to as efficiency in this work. This efficiency (η) represents the relationship between the output power (P_{out}) delivered to the load and the input power (P_{in}) required from the source. It is quantified as the ratio of output power to input power (2.26) and can be alternatively calculated from the total losses using (2.27).

$$\eta = \frac{P_{out}}{P_{in}}. \tag{2.26}$$

$$= \frac{P_{out}}{P_{out} + P_{loss}}. \tag{2.27}$$

Furthermore, the power density (P_{dens}) is defined, which indicates the spatial efficiency of the conversion process. It is described by the ratio of output power to active converter area (A_{conv}), as explicated by

$$P_{dens} = \frac{P_{out}}{A_{conv}}. \tag{2.28}$$

2.2.4.2 Relationship

Figure 2.6 shows the efficiency versus power density of an optimized two-phase converter [10, 30]. Two distinct regimes emerge based on power density: the low-power-density regime and the high-power-density regime. This distinction is attributed to the prevalence of different loss contributors in each regime.

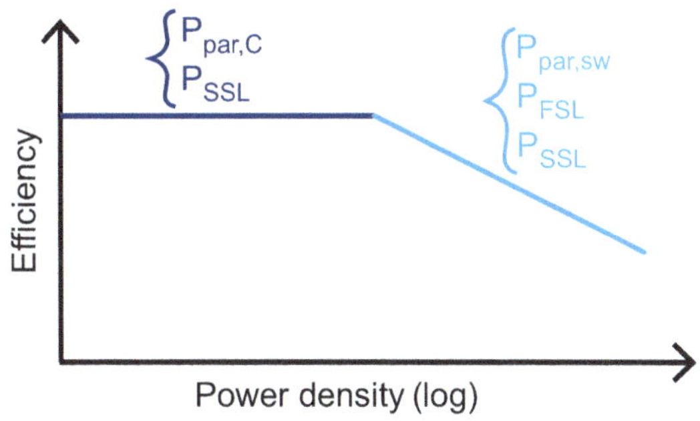

Fig. 2.6 Efficiency versus power density in an optimized two-phase SCC. The dominant loss contributors are annotated in each regime

Within the low-power-density regime, primary losses stem from the flying capacitors, specifically P_{SSL} and $P_{par,C}$. These losses establish an efficiency ceiling, constrained by the ratio of parasitic to useful capacitance of the flying capacitors. Note that advanced multi-phasing techniques [9] can enhance this ceiling to the fundamental limit of leakage. However, these techniques are not suitable for high voltage, as will be discussed in Sect. 3.3.1.

Transitioning to higher power densities, the optimized switching frequency and transistor width both increase with the output current to alleviate series losses (2.14), (2.17). While $P_{par,C}$ is only proportional to this switching frequency (2.24), $P_{par,sw}$ scales linearly with both this switching frequency and this width (2.23). The latter loss component begins to overshadow the former at a specific power density, marking the onset of the high-power-density regime. Within this regime, the switch-related losses and P_{SSL} are dominant. Additionally, as $P_{par,sw}$ escalates more rapidly than the output power in this regime, efficiency diminishes with power density. This introduces a trade-off between efficiency and power density in SCCs.

The transition point between these two regimes hinges on the magnitude of losses in each regime, influenced by the technology, input and output voltages, and the topology. For optimal performance, a converter should operate within the high-power-density regime or in the transitional region. Operating in the low-power-density regime, especially far from the intermediate region, is suboptimal. This is because progressing toward higher power densities does not incur an efficiency penalty, yet it offers a reduced area for a given output power.

2.3 Control

In contrast to the constant input voltage and output current in the previous section, these parameters often vary in real-world applications. Such variations significantly influence the output voltage, as described by (2.6), which is repeated here for convenience:

$$V_{out} = V_{in} \cdot iVCR - I_{out} \cdot R_{out}. \tag{2.29}$$

To maintain a stable output voltage despite fluctuating output current and input voltage, the converter must incorporate load and line regulation. As suggested by (2.29), this can be achieved by manipulating the output impedance or the topology iVCR (Fig. 2.7). By modifying the output impedance, the converter can manage the voltage drop across it, thereby obtaining VCRs lower than the topology iVCR. On the other hand, the iVCR can be altered by changing the topology. The subsequent discussion delves deeper into these two control strategies and elucidates how their combined application enables a wide input voltage range with high efficiency.

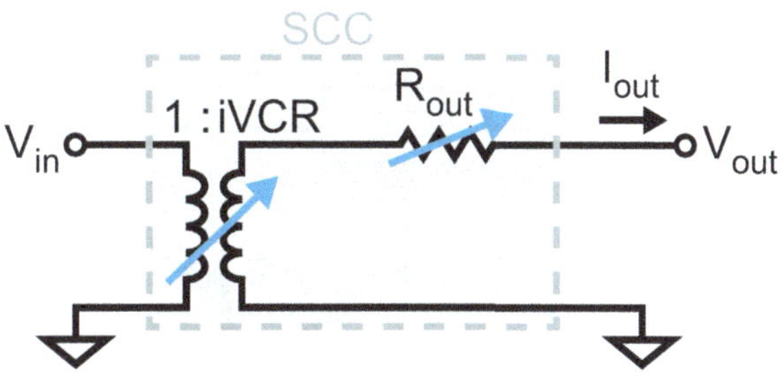

Fig. 2.7 Output voltage regulation by modulating $iVCR$ and R_{out}

2.3.1 Output Impedance Modulation

The first control strategy modulates the output impedance. This output impedance can be varied through different control parameters in both the SSL and FSL regions. In the SSL region, the output impedance can be manipulated through the switching frequency [81] and the flying capacitance [55] (2.14). Alternatively, in the FSL region, it can be controlled by the power switch on-resistance [21] and by the duty cycle through pulse width modulation (PWM) [71] (2.17). This work narrows the focus to the most common approach, which modulates the switching frequency using pulse frequency modulation (PFM) in discrete time.

The widespread adoption of PFM in discrete time is attributed to several advantages [81]. First, it is inherently stable, thereby simplifying the implementation. This contrasts with PWM, which requires a complex frequency compensation due to its poles that vary with the operation point and the load current. Second, in PFM, the switching frequency scales proportionally to the load current, minimizing parasitic switching losses at low output powers (2.22). This enables superior light-load efficiency compared to alternative control parameters. Third, discrete-time PFM can instantaneously alter the effective switching frequency, facilitating rapid response to load changes. Despite these benefits, PFM suffers from an unpredictable noise spectrum, posing challenges in adhering to electromagnetic interference (EMI) standards.

Figure 2.8 illustrates the operational principle of a discrete-time PFM, implemented with a lower-bound clocked hysteretic controller. A clocked comparator assesses the output voltage against a reference voltage (V_{ref}). When V_{out} drops below V_{ref} due to the load current, the comparator initiates a charge event (CE) at the ensuing rising clock edge ($f_{clk,out}$). This event alters the phase of the SCC, which transfers charge to the output so that V_{out} exceeds V_{ref} again. The SCC then remains idle until V_{out} falls once more below V_{ref}. As observed in Fig. 2.8, an increase in the output current causes the charge events to occur more frequently. This dynamically adjusts the switching frequency of CE, effectively modulating the output impedance as intended.

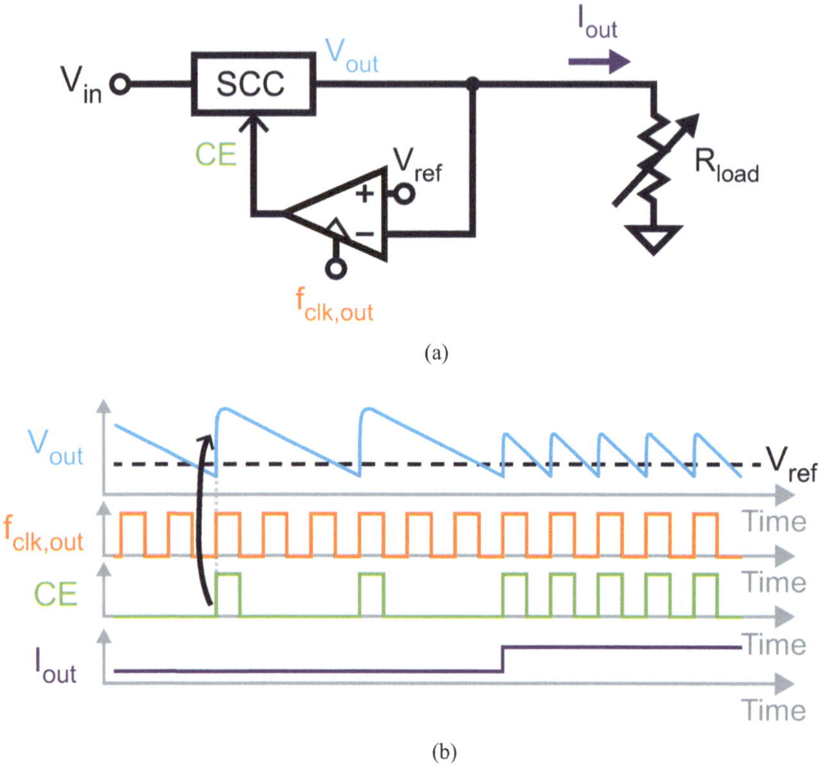

Fig. 2.8 Working principle of a lower-bound clocked hysteretic controller. (**a**) Schematic. (**b**) Waveforms

2.3.2 *iVCR Modulation*

While output impedance modulation is suitable for small input voltage variations, its efficiency markedly diminishes for a wide input voltage range. This reduction in efficiency stems from the dissipation of excess voltage in the output impedance, mirroring the behavior of a linear regulator. To quantify this, the ideal efficiency (η_{ideal}) is calculated using the equivalent model in Fig. 2.2, considering exclusively intrinsic losses:

$$\eta_{ideal} = \frac{P_{out}}{P_{in}}, \tag{2.30}$$

$$= \frac{V_{out} \cdot I_{out}}{V_{in} \cdot iVCR \cdot I_{out}}, \tag{2.31}$$

$$= \frac{VCR}{iVCR}. \tag{2.32}$$

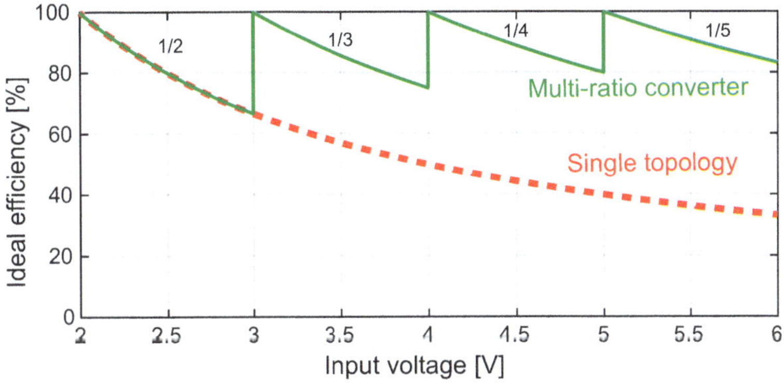

Fig. 2.9 Ideal efficiency versus DC input voltage for a single topology and a multi-ratio converter, with iVCRs annotated

Equation (2.32) confirms that high efficiency is only attainable when the actual voltage conversion ratio closely matches the iVCR of the topology.

To maintain proximity to the iVCR across a wide range of input voltages, multiple iVCRs are realized. This approach combines multiple topologies, each with a distinct iVCR, within a single converter—a concept known as a gearbox or multi-ratio converter. The discrete nature of iVCR adjustments requires output impedance modulation for a continuous transition between neighboring iVCRs.

The efficiency improvement of a multi-ratio converter over a single-ratio topology is illustrated in Fig. 2.9, which plots their ideal efficiency against input voltage. The single-ratio topology is fixed at an iVCR of $1/2$, whereas the multi-ratio converter includes the additional iVCRs of $1/3$, $1/4$, and $1/5$. For the single-ratio topology, efficiency plummets at higher input voltages, as the VCR diverges from the iVCR. Conversely, the multi-ratio converter maintains high efficiency throughout the entire input range, as it can select the most efficient topology for each specific input voltage.

Incorporating more iVCRs enhances efficiency around their corresponding input voltages, but it also introduces several drawbacks. Primarily, adding extra configuration switches to reconfigure the topologies results in the introduction of parasitic resistors and capacitors, which adversely affect the overall efficiency. Another disadvantage is that it increases the system complexity. These two factors collectively constrain the number of iVCRs in a design.

2.4 Conclusion

SCCs transfer charge from the input to the output by periodically changing the configuration of their flying capacitors. Charging and discharging the flying capacitors requires a voltage drop proportional to the output current, represented by a resistive output impedance in an equivalent model. Besides the intrinsic losses linked to this output impedance, extrinsic losses occur due to the periodic charging of parasitic capacitors present in the integrated components.

The SCC control ensures a consistent output voltage, irrespective of a varying load and input voltage. To achieve high efficiency over a wide range of input voltages, SCCs employ gearbox or multi-ratio converters. These converters merge multiple topologies, each characterized by a unique iVCR. To span the continuous range between these discrete iVCRs, the output impedance is modulated between iVCRs. This modulation is typically controlled by fine-tuning the switching frequency through discrete-time PFM.

This chapter laid the groundwork for comprehending the proposed topologies and implementations in the upcoming chapters. Understanding the losses in SSCs is crucial to developing a high-voltage SCC with a good trade-off between efficiency and power density, which will be explored in the following chapter.

Chapter 3
Fully Integrating a High-Voltage DC-DC Converter

Chapter 1 motivated the suitability of switched-capacitor converters (SCCs) for achieving a high input voltage ($\sim$400 V) and complete integration within the milliwatt power range. Therefore, in Chap. 2, the fundamentals of SCCs were elucidated. However, state-of-the-art fully integrated SCCs are confined to a maximum input voltage of 42 V [15]. This chapter targets to push this maximum input voltage to 400 V DC. The proposed topology overcomes the key challenges associated with the increasing input voltage, which include diminished component quality and heightened parasitic losses.

This chapter starts by elaborating on the specifications and discussing the shortcomings of state-of-the-art approaches. The following part examines the key challenges at high input voltages and proposes a topology to tackle these challenges. Subsequently, the implementation details are covered, including the customization of high-voltage capacitors and drivers, as well as an improved control circuit. Finally, Sect. 3.6 presents the measurement results and compares them with the state of the art.

Parts of this chapter were presented at the Custom Integrated Circuits Conference (CICC) of 2022 in Newport Beach, USA [83], and have been published in the Journal of Solid-State Circuits (JSSC) in 2023 [85]. Additionally, a portion of this chapter is accepted for publication in the Advances in Analog Circuit Design (AACD) book in 2023 [87].

3.1 Specifications

Chapter 1 briefly introduced the specifications of the voltage converter. This section assigns specific values and priorities for both functional and performance specifications of the DC-DC converter in this chapter. While different applications prioritize other specifications, the focus in the first two subsections lies on universally critical

© The Author(s), under exclusive license to Springer Nature Switzerland AG 2025

T. Van Daele, F. Tavernier, *Fully Integrated High-Voltage DC-DC and AC-DC Conversion*, Analog Circuits and Signal Processing,
https://doi.org/10.1007/978-3-031-78519-1_3

specifications. Subsequently, Sect. 3.1.3 enumerates specifications contingent on the specific application and those essential for an industrial context. Note that the given specifications are also applicable to the converters in the subsequent chapters, except for the input voltage requirement.

3.1.1 Functionality Specifications

Functionality specifications, as suggested by their name, are crucial for realizing the intended functionality. In particular, the converter should bridge the significant voltage gap between available high-voltage sources and emerging low-power applications, as previously described in Sect. 1.1. This requirement translates into the following three functional specifications:

- **Input voltage**
  The input voltage is DC and is set at 400 V, in line with the high bus voltage commonly found in electric cars and industrial environments. Furthermore, it surpasses the peak voltage of the AC mains worldwide, allowing for the extension of this chapter's DC-DC converter to AC mains inputs in subsequent chapters.
- **Output voltage**
  The output voltage is set at 12 V DC, serving as an intermediate voltage to power chips with supply voltages of $\leq$5 V DC. Multiple output voltages are typically required for the different specific loads such as digital, input/output (I/O), and analog. Further segmenting these supply voltages can optimize them for their specific load, enhancing efficiency. Rather than generating all these voltages within the high-voltage converter, it is beneficial to use a dedicated second stage. This second stage operates at a lower voltage, which is better suited for implementing this complexity. Additionally, the 12 V DC matches the requirements of an LED driver, which typically operates at 12 V DC or 24 V DC [62], where several LED units are put in series.
- **Output power**
  The output power spans the milliwatt range, reaching a maximum of a few hundred milliwatts. This aligns with the intended applications, covering home automation, grid infrastructure, LEDs, and control circuitry [45, 76].

3.1.2 Performance Specifications

While functionality specifications are fundamental for attaining the target functionality, performance specifications reflect the quality of the execution. Consequently, rather than establishing specific values, it is more meaningful to arrange them based on their relative importance. The subsequent specifications are enumerated in descending order of priority for this work:

- **Fully integrated**
  Fully integrating the converter yields a cost-effective and compact solution that enhances the performance across various specifications. Additional details on this subject were elaborated upon in Sect. 1.2.
- **Power density**
  The power density determines the chip area required to achieve a specified output power. The chip area directly impacts the fabrication cost and, consequently, should be minimized. Since in an integrated solution, the area can be reduced by scaling down the output power, typically the power density is used as a metric instead of the chip area or the output power.
- **Efficiency**
  High efficiency translates to low losses, thereby improving the battery life. Moreover, these losses also influence the effort and cost to dissipate heat, especially for high on-chip power levels. Efficiency is assigned a lower priority, provided that the associated losses do not require expensive cooling systems and packaging.

While both power density and efficiency are important, there exists a trade-off between them (cf. Sect. 2.2.4.2), with the optimal balance depending on the specific application. This work prioritizes enhancing the power density, even at the cost of a slight efficiency reduction. Such a trade-off is advantageous in applications where the converter power losses are minor compared to the overall system losses, as long as heat dissipation is not critical.

In cars, for example, electronics consume only a fraction of the power required for driving [90]. As a result, most voltage converters used at low voltage are linear regulators, despite their low efficiency. However, the emerging high input voltages lead to a drastic drop in efficiency for linear regulators, posing thermal challenges. Consequently, linear regulators get replaced by more efficient switching converters. Within the domain of switching converters, priority is once again given to power density over minor efficiency gains in the target power range, given that their efficiency typically surpasses the thermal requirements.

3.1.3 Other Specifications

Other important specifications include line and load regulation, output-voltage ripple, input voltage range, and quiescent power consumption. Since these parameters are highly dependent on the specific application, they are not discussed in detail.

Galvanic isolation is generally not a requirement for targeted applications such as sensors, RF transmitters, and LED strings with standalone operation [59]. While galvanic isolation is indispensable for certain applications to guarantee safety compliance, it is unnecessary for standalone operation without an electric connection to the environment. This eliminates the possibility of contact with the human body, mitigating the risk of electric shocks.

In an industrial setting, additional constraints like electrostatic discharge (ESD), electromagnetic interference (EMI), and protection circuitry (e.g., short circuit protection) must also be considered. However, these considerations fall beyond the scope of this work.

3.2 State of the Art and This Work

The primary objective of the DC-DC converter is to achieve a high input voltage and a high power density while being fully integrated. This work focuses on power density rather than output power, given that output power can be augmented by simply expanding the converter area. Furthermore, the large input voltage implies a large voltage conversion step, given the low output voltage.[1]

A linear regulator can meet the integration specification and tolerates high input voltages of 450 V [44]. However, it is unsuitable due to the maximum 3% efficiency for the large conversion step, as the excess voltage is entirely dissipated in the pass device. For instance, assuming an output power of 200 mW, this results in a power loss of 6.4 W. Dissipating the generated heat dictates a big package, thereby limiting the power density and cost efficiency.

Alternatively, conventional approaches such as flyback topologies [12, 13], multilevel converters [89, 94, 95], flybuck converters [19, 20], and soft-switching converters [6, 16, 91] need a bulky inductor or transformer, making it spacious and costly [59]. Although recent research [59] reduces the number of external components to only two, the bulky inductor cannot be integrated. Furthermore, a large conversion step typically requires extremely low duty cycles, limiting higher switching frequencies required for complete integration and challenging control (cf. Sect. 1.3.2). Another approach [22, 43] uses a discrete switched-capacitor topology that eliminates the need for inductors and transformers. However, it inherently relies on numerous external flying capacitors to achieve a large conversion step [37].

As opposed to the approaches above, a fully integrated switched-capacitor converter can circumvent the need for both external components and low duty cycles. It exclusively uses switches and capacitors, enabling a fully integrated solution in CMOS. The topology sets the conversion step independent of the duty cycle, which enables larger switching frequencies to scale down the passives toward integration.

Although fully integrated SCCs are promising for realizing big conversion steps, the state-of-the-art solutions can only handle input voltages up to 42 V [15], as shown in Fig. 3.1. Furthermore, SCC performance becomes worse with

[1] Note that this chapter uses the term ideal conversion step (N) instead of ideal voltage conversion ratio (iVCR). This terminology avoids confusion when referring to large conversion steps. This is in contrast to previous work that used the term large conversion ratios to denote these substantial conversion steps. However, this is misleading as, mathematically, a large conversion ratio implies a lower conversion step (2.5).

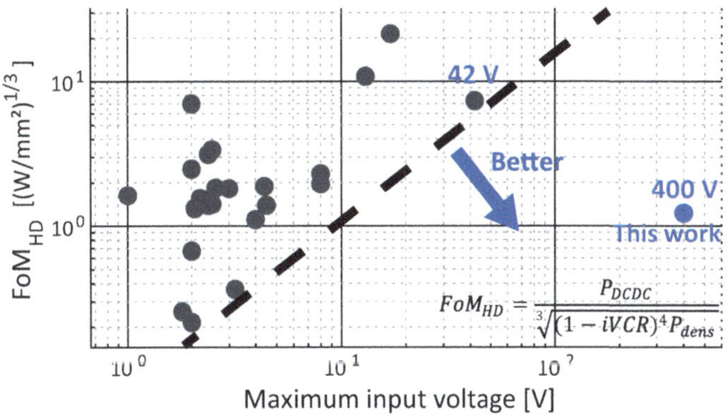

Fig. 3.1 Performance versus maximum input voltage for state-of-the-art fully integrated DC-DC converters, based on [69]

larger input voltages due to the decreased component quality and the increased parasitic losses, as will be explained in Sect. 3.3. The performance is quantified by the high-density figure of merit (FoM_{HD}) [10], described by (3.1). This metric considers the ideal voltage conversion step (N), the power density (P_{dens}), and the normalized converter losses (P_{DCDC}).[2] Including these three parameters ensures a fair comparison, given their interdependence. A lower FoM_{HD} value corresponds to better performance.

$$FoM_{HD} = \frac{P_{DCDC}}{\sqrt[3]{(1 - N^{-1})^4 \cdot P_{dens}}}. \tag{3.1}$$

This work presents a fully integrated SCC with an input voltage of 400 V, more than 9× higher than prior fully integrated DC-DC converters. While in the state-of-the-art (Fig. 3.1) higher input voltages diminish the performance, this work achieves a good trade-off between power density and efficiency through innovative topology techniques ($FoM_{HD} = 1.23$).

3.3 Challenges

To determine the most suitable topology for achieving complete integration at high input voltages, one should first understand the primary challenges. Section 3.3.1

[2] To simplify the comparison with other works at this stage, this figure of merit focuses on efficiency and power density. These two parameters are important for a wide range of applications [8]. Nevertheless, note that when using a figure of merit, a considerable amount of information gets lost. In particular, other parameters like output power may be overlooked, yet they are essential in some applications.

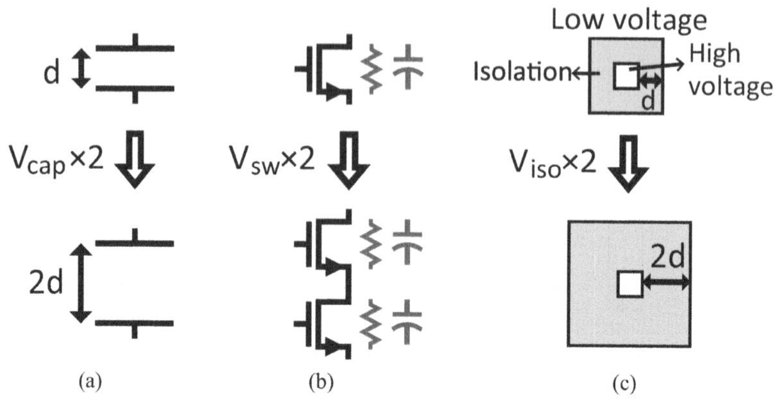

Fig. 3.2 Impact of increasing the voltage rating of (**a**) capacitors, (**b**) switches, and (**c**) isolation

investigates the factors contributing to the decrease in component quality. Subsequently, the impact of high input voltages on parasitic losses is explored. Finally, Sect. 3.3.3 elucidates how the large technology node, required to accommodate high voltages, restricts the power density.

3.3.1 Reduced Component Quality

The high-voltage capability compromises the quality of the components. For instance, increasing the voltage rating of a capacitor (V_{cap}) requires a larger distance between its plates (d) to prevent exceeding the breakdown electric field (Fig. 3.2a). This larger spacing results in lower capacitance compared to its low-voltage counterpart, which leads to higher losses as described in (2.14) from the previous chapter. Similarly, when the voltage rating of a transistor (V_{sw}) increases, the performance deteriorates. For example, the switch in Fig. 3.2b has on-resistance and parasitic capacitances. When two such switches are stacked (i.e., cascoded) to double the voltage rating, both the on-resistance and parasitic capacitances also double. This results in more power losses (2.17), (2.23). Section 3.4.1 provides a more in-depth discussion of how the quality of capacitors and switches scales with the voltage rating.

Furthermore, increasing the voltage of a component relative to its low-voltage surroundings requires a greater isolation distance d (Fig. 3.2c). This isolation is required to avoid exceeding the breakdown voltage between the component and both its neighboring components and the substrate. As depicted in Fig. 3.3, there are two methods to implement this isolation [1]: using a reverse-biased p-n junction or employing a dielectric material like an oxide. Junction isolation needs an enormous area for high-voltage isolation because the depletion region expands with increasing voltage [28]. In contrast, dielectric isolation provides more compact high-

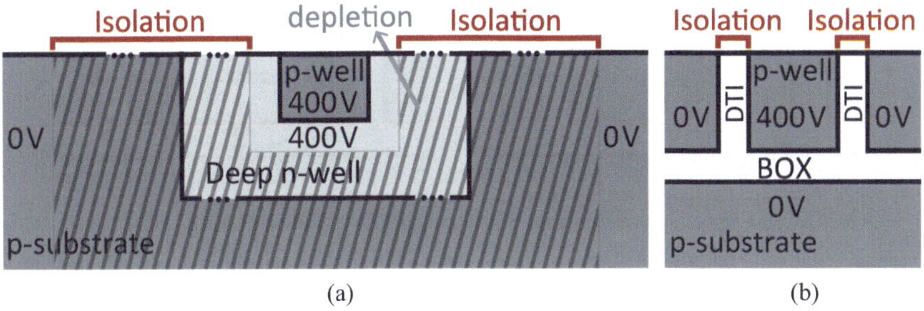

Fig. 3.3 Isolation methods. (**a**) Reverse-biased p-n junction. (**b**) Dielectric material

voltage isolation and is available in silicon-on-insulator (SOI) technologies. For example, Fig. 3.3b illustrates an SOI technology where a p-well at high voltage is isolated from the low-voltage surroundings. The buried oxide (BOX) avoids vertical breakdown to the substrate (= handle wafer), while deep trench isolation (DTI) isolates the p-well laterally from neighboring components. The converters in this work are implemented using such an SOI technology, considering its highly compact isolation at 400 V. This choice results in a significant reduction in chip area, outweighing their higher cost per area.[3]

Although SOI technology greatly reduces the isolation area, this area remains relatively large for small components. As the component size shrinks, the isolation area becomes dominant due to its fixed distance around the component, as illustrated in Fig. 3.2c. This restricts the segmentation of components into smaller units that handle a proportional fraction of the power. Consequently, advanced multi-phasing techniques [9, 10], which rely on the costless segmentation of components in a low-voltage integrated solution, become unsuitable for high input voltages [33].

Compared to low-voltage components, the isolation area overhead is substantially worse for high-voltage transistors (diodes), which block their high voltage between their drain (cathode) and source (anode). Before examining this in further detail, first consider a low-voltage transistor that floats at high voltage within an isolated p-well, as depicted in Fig. 3.4a. In this case, the electric field lines of the transistor are entirely confined within its p-well. In contrast, a high-voltage transistor's electric field is strong enough to extend into adjacent silicon areas, as shown in Fig. 3.4b. The electric field lines of the high-voltage transistor interact with surrounding structures such as the conductive substrate, causing the electric field to compress. This compression locally increases the electric field beyond the critical field of the oxide or the silicon, resulting in breakdown. To prevent this, high-voltage transistors require field termination structures that spread out the electric field. Examples of such structures include DTI structures and reverse-biased handle-

[3] Additional benefits of SOI technologies include their reduced parasitic capacitances, their availability at 400 V, and the elimination of parasitic bipolar effects to the substrate [93]. The latter enhances reliability and simplifies design complexity.

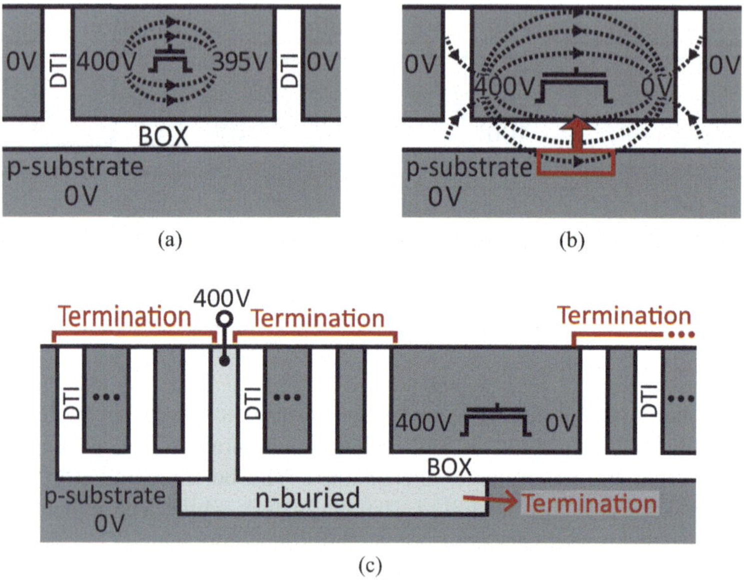

Fig. 3.4 (**a**) Low-voltage transistor floating at high voltage and its associated field lines. (**b**) High-voltage transistor exhibiting compressed electric field lines. (**c**) High-voltage transistor with field termination structures

wafer diodes [23, 59], as illustrated in Fig. 3.4c. The reverse-biased handle-wafer diode is formed by implementing an n-buried layer into the p-substrate (= handle wafer) just beneath the buried oxide layer.

Note that high-voltage capacitors exhibit significantly less isolation area overhead compared to these high-voltage transistors and diodes. These capacitors are implemented as sandwich capacitors, confining the electric field within the layers of the metal stack. This prevents field compression by surrounding structures, eliminating the need for elaborate field termination structures.

3.3.2 Increased Parasitic Losses

The second challenge is the increased parasitic switching losses at high input voltages. These parasitic losses (P_{par}) occur due to the periodic charging of the parasitic capacitors of the components (C_{par}) by their voltage swing (ΔV_{par}) at the switching frequency (f_{sw}). This is described by (2.22), repeated here for convenience:

$$P_{par} = f_{sw} \cdot C_{par} \cdot \Delta V_{par}^2. \tag{3.2}$$

All three variables are impacted by high input voltages. High input voltages are generally associated with substantial voltage swings across the parasitic capacitors. This leads to significant losses for a given C_{par} and f_{sw}, because of the quadratic relationship between the parasitic losses and the voltage swings (3.2). Furthermore, when capacitors are blocking high voltage, their capacitance density is low, as previously explained in Sect. 3.3.1. To counteract this, a higher switching frequency and/or capacitor area ($\propto C_{par}$) is required, thereby increasing the parasitic losses as well (3.2). Additionally, SCCs inherently require numerous components to achieve large voltage conversion steps [37]. Since these components are integrated close together, the mutual coupling between the adjacent components becomes substantial, raising C_{par} and, in turn, P_{par} (3.2).

3.3.3 Large Technology Node

The third challenge stems from the limitation that high-voltage capabilities are only provided in large technology nodes, as was discussed in Sect. 1.5. For instance, X-FAB offers 400 V down to 180 nm [93], GlobalFoundries 700 V in 180 nm [50], and ST-Microelectronics 200 V in 160 nm [70]. Larger technology nodes inherently come with larger parasitic capacitances, which, in turn, impede the speed of the digital control loop. This limits the maximum switching frequency of the SCC, ultimately constraining the power density.

3.4 Topology

To address the three challenges posed by high input voltages, a topology is presented. Initially, Sect. 3.4.1 investigates how components should block the high voltage to mitigate the diminished component quality. Next, the sidewall coupling reduction technique is proposed to counteract the increased parasitic losses. Finally, Sect. 3.4.3 enhances the power density despite the limited switching frequency inherent to high-voltage technologies.

3.4.1 Blocking the High Voltage

This section quantizes the component performance versus voltage rating of the primary components in switched-capacitor converters. Using the derived equations, different topologies for blocking high input voltages are compared based on their losses for the same power density.

3.4.1.1 Switch Performance Versus Voltage Rating

The performance of a switch is defined by the product of its on-resistance (R_{on}) and its switching energy to drive its gate and drain capacitors (E_{sw}) [2]. Equation (3.3) expresses the switch performance versus its voltage rating (V_{sw}) by the exponent α_s:

$$R_{on} \cdot E_{sw} \propto V_{sw}^{\alpha_s}. \tag{3.3}$$

The actual scaling behavior is complex and depends on the switch type for a given voltage rating and the technology. To theoretically compare topologies, prior studies [26, 49, 59, 66, 68] derived an α_s between 2 and 3 for silicon-based power switches under a constant-electric-field scaling. A cubic relationship ($\alpha_s = 3$) is more accurate for single devices where the on-resistance is dominated by the channel, which is usually not the case for high-voltage switches [26, 68]. On the other hand, quadratic scaling ($\alpha_s = 2$) is more precise when low-voltage devices are cascoded to achieve a higher voltage rating [49, 66].

For high-voltage switches, $\alpha_s = 2$ is a reasonable approximation, assuming that the switch-parasitic losses are predominantly related to the drain-source capacitance [59]. However, this is a best-case scenario, as the losses from other parasitic capacitors (such as the substrate capacitor and capacitors to adjacent wells) are scaling less favorably with the voltage rating. Nevertheless, an $\alpha_s = 2$ will be utilized for the topology comparison later in this section. The implications of a higher α_s will be addressed at the end of this comparison.

It is worth mentioning that wide-bandgap materials like gallium nitride (GaN) and silicon carbide (SiC) ameliorate the switch scaling trend with the voltage rating, which results in $1 < \alpha_s < 2$ [26, 68]. However, their lack of suitable pMOS devices limits the switching frequency and challenges the integration of digital circuitry and drivers on the same substrate, as was discussed in Sect. 1.5.

3.4.1.2 Capacitor Performance Versus Voltage Rating

This subsection analyzes the relationship between the capacitor voltage rating and the capacitance density. As the voltage rating rises, capacitors need a thicker dielectric between their plates to withstand the stronger electric field and avoid breakdown. This thicker dielectric results in a lower capacitance density compared to low-voltage counterparts.

To deduce the scaling trend, previous works [8, 66] assumed that all capacitor types occupy the same dielectric volume under a maximum electric field, regardless of their voltage rating. The maximum electric field means that the dielectric thickness between the capacitor plates is chosen almost to reach breakdown, to optimize the energy density for every voltage rating. This assumption of constant dielectric volume results in a constant area-specific energy density ($E_{dens,A}$), as both the dielectric volume (Vol_d) and the electric field (E_F) remain constant in the

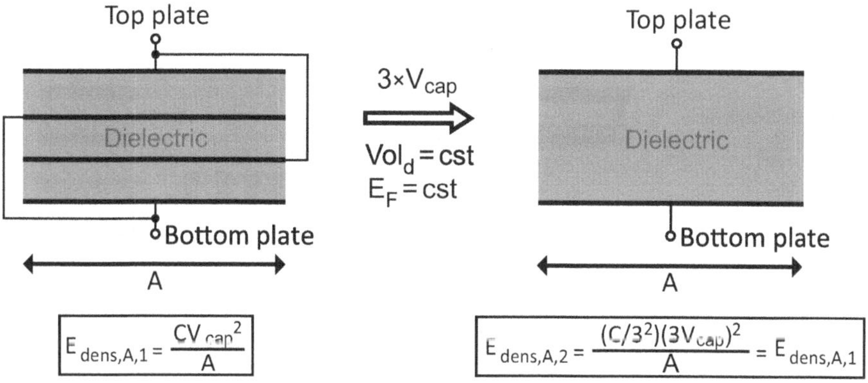

Fig. 3.5 Comparison of the area-specific energy density of capacitors when increasing the voltage rating (V_{cap}). The dielectric volume (Vol_d) and electric field (E_F) are assumed constant (cst)

following formula [8]:

$$E_{dens,A} = \frac{\epsilon \cdot E_F^2}{2 \cdot A} \cdot Vol_d, \tag{3.4}$$

where ϵ is the dielectric permittivity and A the area.[4]

This constant area-specific energy density, assuming the fixed dielectric volume, is illustrated in Fig. 3.5. When the voltage rating is increased by $3\times$, the dielectric thickness must be raised by the same factor to maintain a constant electric field. This greater distance between the capacitor plates reduces the capacitance by $3\times$. Furthermore, only a single sub-capacitor can now fit in the dielectric in Fig. 3.5, resulting in another $3\times$ reduction in capacitance. Consequently, the area-specific energy density remains unchanged compared to the low-voltage capacitor, as depicted in Fig. 3.5.

The constant area-specific energy density in the earlier works leads to a quadratic reduction in capacitance density with the increasing capacitor voltage rating (V_{cap}). This stems from the equation for the energy stored on a capacitor for a specified area:

$$E_{dens,A} = \frac{C \cdot V_{cap}^2}{2 \cdot A}. \tag{3.5}$$

However, the assumption of a constant dielectric volume under the maximum electric field in prior works does not reflect the actual situation. In reality, blocking

[4] Note that the area-specific energy density remains unaffected by the value of ϵ when operating under a maximum electric field. The dielectric breakdown strength decreases inversely with the square root of ϵ [38, 39], leading to a constant $E_{dens,A}$ (3.4).

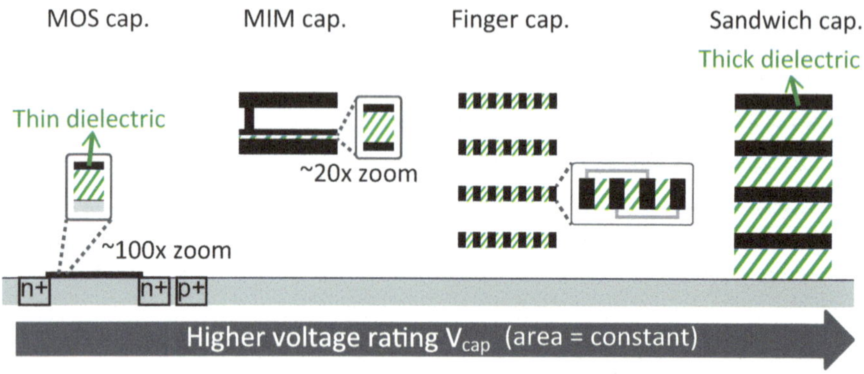

Fig. 3.6 Integrated capacitor types with low (left) to high (right) voltage ratings (V_{cap}). Their dielectric volume, operating at the maximum electric field, is highlighted

a higher voltage in a standard integrated technology demands different capacitor types with a thicker dielectric than their low-voltage counterparts, as illustrated in Fig. 3.6.

More specifically, as the dielectric thickness increases, capacitor types transition from MOS to metal-insulator-metal (MIM) to finger (or vertical parallel plate) to sandwich capacitors. The green hedged volumes in Fig. 3.6 illustrate that high-voltage capacitors utilize a larger dielectric volume at their maximum electric field by exploiting the available height. For example, the sandwich capacitor uses a large dielectric volume at high blocking voltages, operating at its maximum electric field. In contrast, at low voltages, the MOS capacitor is used, which only occupies a small portion of the available volume at the maximum electric field. This is because, at low blocking voltages, the electric field in the thick dielectric of the sandwich capacitor is weak. As a result, stacking a sandwich capacitor on top of the MOS capacitor would have a minimal impact on the overall energy density (3.4).

To model the beneficial effect of the increased dielectric energy density at high voltage, Fig. 3.7 illustrates the relationship between the capacitance density (C/A) and the capacitor voltage rating V_{cap} for commercial high-voltage technologies [48]. The dotted lines represent the inverse linear and quadratic scaling trends as a reference. From this graph, one can derive the following empirical formula:

$$\left(\frac{C}{A}\right)_{empirical} \propto \frac{1}{V_{cap}}. \tag{3.6}$$

Equation (3.6) is more accurate than the quadratic relation assumed in previous works, which confirms the increased dielectric energy storage at high voltage.

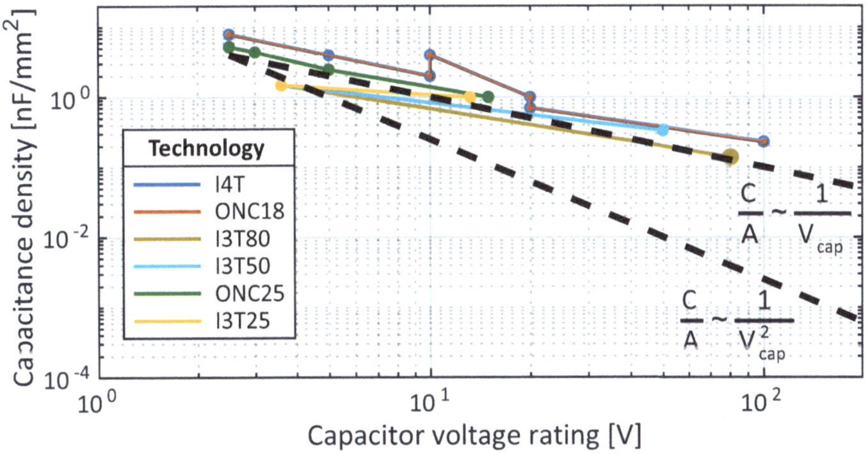

Fig. 3.7 Scaling trend of capacitance density (C/A) with capacitor voltage rating (V_{cap}) for commercial high-voltage CMOS technologies [48]

3.4.1.3 Comparison

The component performance equations versus voltage rating enable a comparison of SCC topologies at high input voltage. These topologies are depicted in Fig. 3.8, where V_{in}, V_{out}, and q_{out} represent the input voltage, output voltage, and output charge, respectively. Three options are available to block the high input voltage at large ideal voltage conversion steps (N). First, high-voltage power switches can bridge the large voltage gap, decreasing the switch quality quadratically as previously discussed (e.g., series-parallel topology in Fig. 3.8c). In the second option, the flying capacitors block the high voltage (e.g., Dickson-star topology in Fig. 3.8b). Lastly, the third option stacks low-voltage flying capacitors and low-voltage power switches (e.g., Dickson-ladder topology in Fig. 3.8c). While the low-voltage components offer superior quality, they must transfer more charge than the other topologies to achieve the same output power, which increases the output impedance losses.

To determine the optimal topology for large N at high voltage, a loss analysis compares their losses for a given area. Prior research [40] has also conducted a loss analysis specifically for large N but still holds some caveats. First, it does not consider the impact of the voltage rating on the component performance in the comparison metrics, which is crucial at large N and high voltage. Second, it neglects both the FSL and parasitic losses of the switches, rendering it unusable for high power densities. Another study [8] provides a general loss analysis for switched-capacitor topologies, but it assumes that capacitors have a cost proportional to their voltage rating squared, changing the performance results.

To get a fair comparison between the topologies, the losses of each topology are optimized to all the design variables, while maintaining the same power density. In

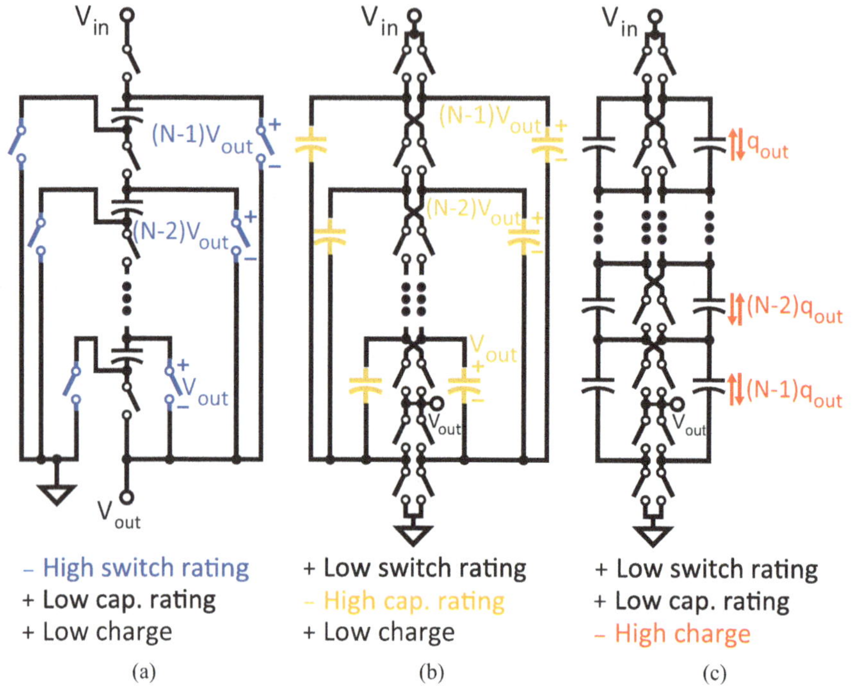

Fig. 3.8 Topological options to block the high input voltages toward large ideal voltage conversion steps (N): (**a**) high-voltage power switches (series-parallel), (**b**) high-voltage capacitors (symmetrical Dickson-star), and (**c**) stacked low-voltage components (symmetrical Dickson-ladder). The blocking voltages/charges that significantly impact the losses are annotated in color

SCCs, different loss mechanisms dominate depending on the power density, as was covered in Sect. 2.2.4.2. In the low-power-density regime (LD), the predominant losses are those associated with the flying capacitors. On the other hand, switch-related and charge distribution losses dominate in the high-power-density regime (HD) [8, 30]. However, it is insufficient to only show the outcomes for HD, despite the high power density requirement. At high input voltages, the parasitic losses in the flying capacitors (associated with LD) increase, as discussed in Sect. 3.3.2. As a result, the converter still operates in the transitional region between LD and HD for the target power densities, unlike low-voltage designs. Therefore, the losses of both regimes will now be given. Following the method in [8], Eqs. (3.7) and (3.8) provide the optimized size of the kth flying capacitor ($A_{C,k,LD}$) and the optimized normalized losses ($P_{N,LD}$) for the low-power-density regime, respectively.

$$A_{C,k,LD} \propto \frac{a_{c,k} \cdot \sqrt{V_{cap,k}}}{\Delta V_{par,C,k}}. \tag{3.7}$$

$$P_{N,LD} \propto \sum_{caps,i} \left(\sqrt{V_{cap,i}} \cdot |a_{c,i} \cdot \Delta V_{par,C,i}| \right). \tag{3.8}$$

Equation (3.7) provides a guideline for determining the size of the kth flying capacitor based on their charge multiplier element ($a_{c,k}$), flying capacitor blocking voltage (V_{cap}), and the voltage swing of their parasitic capacitors ($\Delta V_{par,C,k}$) for this regime. Equations (3.9), (3.10), and (3.11) provide $A_{C,k,HD}$, the optimized size of the kth power switch ($A_{sw,k,HD}$), and the optimized normalized losses ($P_{N,HD}$) for the high-power-density regime, respectively.

$$A_{C,k,HD} \propto \sqrt{V_{cap,k}} \cdot |a_{c,k}|, \tag{3.9}$$

$$A_{sw,k,HD} \propto V_{sw,k} \cdot |a_{sw,k}|, \tag{3.10}$$

$$P_{N,HD} \propto \left(\sum_{caps,i} \left(\sqrt{V_{cap,i}} \cdot |a_{c,i}| \right) \cdot \sum_{sw,i} |V_{sw,i} \cdot a_{sw,i}| \right)^{\frac{2}{3}}. \tag{3.11}$$

Herein, $V_{sw,k}$ is the switch blocking voltage and $a_{sw,k}$ the charge multiplier element of the kth power switch.

To compare the topologies at the low-power-density regime, Fig. 3.9a illustrates the optimized normalized losses for this regime versus the ideal conversion step for each topology. As N increases, the symmetrical Dickson-star topology outperforms the other topologies because of the increased dielectric storage for high capacitor blocking voltages. Likewise, in the high-power-density regime (Fig. 3.9b), the symmetrical Dickson-star topology exhibits the least losses thanks to the beneficial capacitor performance with increased blocking voltages. For $N \leq 4$, the series-parallel demonstrates a slight advantage because it needs fewer power switches for the same N. Conversely, the symmetrical ladder has the largest losses due to the greater charge flowing through the components than the other topologies. In conclusion, the most efficient topology for large conversion steps blocks the high voltage with capacitors, thereby maximizing the dielectric energy storage. Note that the assumed $\alpha_s = 2$ for the topology comparison is the minimum α_s, as discussed in Sect. 3.4.1.1. However, if $\alpha_s > 2$, the series-parallel losses would be even greater, reinforcing the topology conclusion.

3.4.2 Sidewall Coupling Reduction

The previous section has shown that for high-voltage designs, topologies with high-voltage flying capacitors are preferred. However, the efficiency is still limited by the significant parasitic losses at high input voltages, as discussed in Sect. 3.3.2. To narrow down to the most efficient choice among the topologies with high-

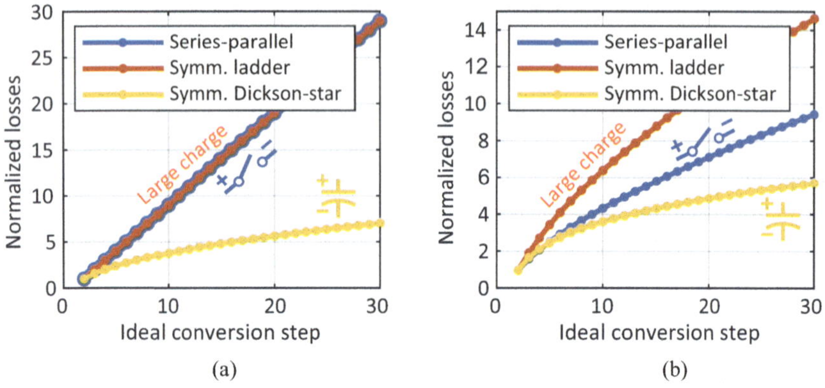

Fig. 3.9 Comparison of normalized losses versus ideal voltage conversion step for high-voltage topologies in both power density regimes. (**a**) Low-power-density regime. (**b**) High-power-density regime

voltage capacitors, this subsection chooses a topology that enables sidewall coupling reduction, which boosts efficiency.

The parasitic losses are primarily generated by the substrate and sidewall capacitance of the flying capacitors. The substrate coupling can be significantly reduced by placing a handle-wafer diode below the flying capacitor [59], as will be elaborated upon in Sect. 3.5.2. However, the efficiency of high step-down SCCs remains constrained by the sidewall coupling between neighboring flying capacitors, as illustrated in Fig. 3.10a. This sidewall coupling stems from the sideways coupling between neighboring flying capacitors (cf. Sect. 2.2.3.2). The inherent requirement for numerous flying capacitors in high step-down SCCs [37] leads to the close integration of neighboring flying capacitors, which causes substantial sidewall coupling. Furthermore, the adjacent flying capacitors in SCCs typically swing in antiphase, which leads to a voltage amplitude across the sidewall capacitor being the sum of the neighboring flying capacitor amplitudes.

To combat sidewall coupling losses (P_{Cside}), this work groups flying capacitors C_{fly} with an identical voltage swing together, as illustrated in Fig. 3.10. When neighboring capacitors share both the same amplitude and phase, the voltage swing (ΔV_{Cside}) across the sidewall capacitor (C_{side}) is eliminated, thereby removing its associated losses. To achieve this, a topology is chosen in which all flying capacitors exhibit voltage swings with equal amplitudes (Fig. 3.10b). However, despite equal amplitudes, a ΔV_{Cside} still exists since adjacent capacitors typically swing in opposite phases in SCC topologies. To address this, Fig. 3.10c places flying capacitors with the same phase together in the layout, eliminating their sidewall coupling. These capacitors, having the same voltage swing amplitude and phase, will now be referred to as a *group*.

The improvement of the sidewall coupling reduction technique can be theoretically derived. Previous studies have demonstrated that the converter efficiency is

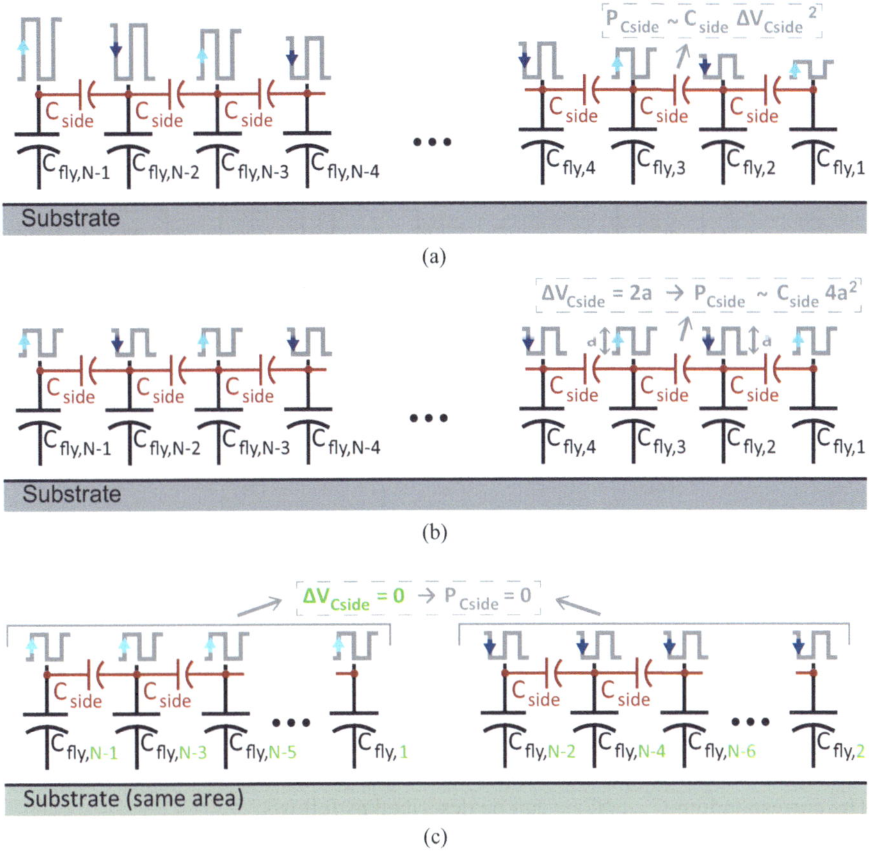

Fig. 3.10 (**a**) Conventional topologies suffer from sidewall coupling loss (P_{Cside}). These losses are eliminated by (**b**) choosing a topology where the flying capacitors have a voltage swing with the same amplitude a, and (**c**) putting the flying capacitors with the same phase together

constrained by the ratio of the parasitic capacitance of the flying capacitor (C_{par}) and the main capacitance of the flying capacitor (C_{fly}) [30] (cf. Sect. 2.2.4.2). In this context, C_{par} is assumed equivalent to C_{side}, given that the sidewall coupling dominates over the substrate coupling. The value of C_{fly} is proportional to the area, while C_{side} and thus C_{par} is proportional to the perimeter. Assuming that all flying capacitors are square-shaped with side length z, as depicted in Fig. 3.11a, the ratio of C_{par} and C_{fly} can be described by the following equation for conventional (conv.) designs:

$$\left(\frac{C_{par}}{C_{fly}}\right)_{conv.} = k_{par} \cdot \frac{\text{perimeter(unit cap.)}}{\text{area(unit cap.)}} = k_{par} \cdot \frac{4}{z}, \tag{3.12}$$

with proportionality constant k_{par}.

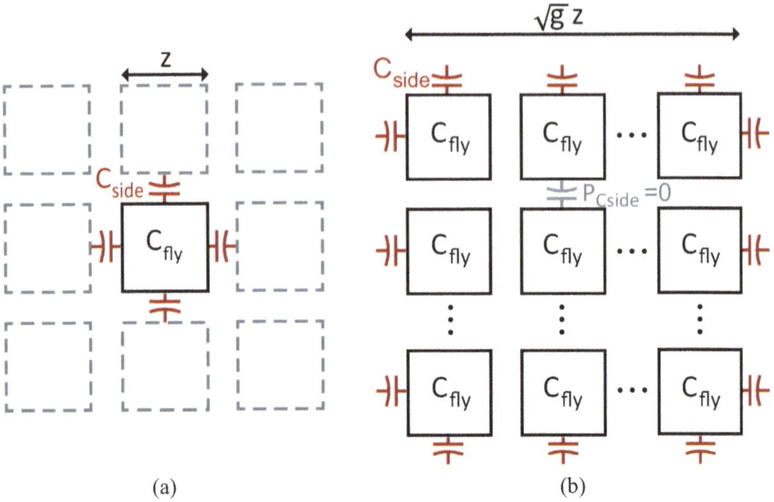

Fig. 3.11 (**a**) Conventionally, all the sidewall capacitors of the flying capacitors generate losses. (**b**) The sidewall coupling reduction technique only has losses at the outer perimeter of an entire group

Figure 3.11b demonstrates that grouping g flying capacitors with identical voltage swing (amplitude and phase) eliminates the losses of their mutual coupling. As a result, the parasitic coupling is only generating losses at the perimeter of the entire group. The overall main capacitance of the group follows from its total area. The corresponding C_{par}/C_{fly} can be described as follows:

$$\left(\frac{C_{par}}{C_{fly}}\right)_{new} = k_{par} \cdot \frac{\text{perimeter(group)}}{\text{area(group)}} = \frac{1}{\sqrt{g}} \cdot \left(\frac{C_{par}}{C_{fly}}\right)_{conv}. \tag{3.13}$$

Therefore, this approach attains a C_{par}/C_{fly} advancement proportional to the square root of the number of flying capacitors clustered together. The impact on the optimized power losses follows from (3.8) and (3.11):[5]

$$P_{N,LD} \propto \sqrt{\frac{C_{par}}{C_{fly}}} \propto \frac{1}{\sqrt[4]{g}}, \tag{3.14}$$

$$P_{N,HD} \text{ is independent of } g. \tag{3.15}$$

[5] Equations (3.8) and (3.11) were not written in full in Sect. 3.4.1 for the sake of clarity. Fully expanding these equations reveals that $P_{N,LD} \propto \sqrt{\frac{C_{par}}{C_{fly}}}$ and that $P_{N,HD}$ is independent of $\frac{C_{par}}{C_{fly}}$.

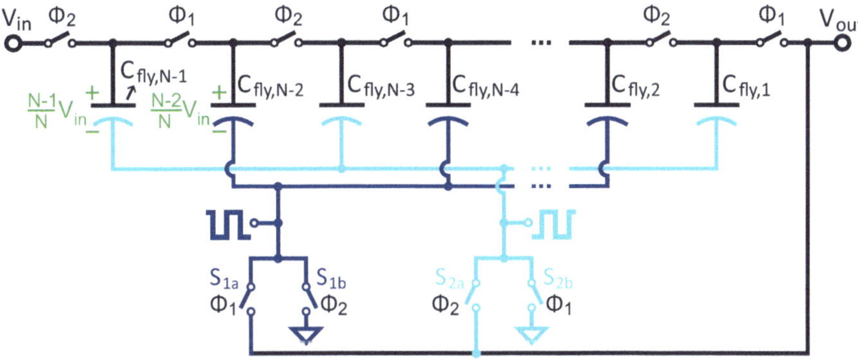

Fig. 3.12 Dickson-star topology with the sidewall coupling reduction technique

The reduction in sidewall coupling only reduces the losses in the LD regime, aligning with the discussion in Sect. 2.2.4.2. However, note that P_{LD} remains significant at high input voltage despite the targeted high power density. This is due to the substantial parasitic losses at high input voltages, as detailed in Sects. 3.3.2 and 3.4.1.

The chosen Dickson-star topology enables the sidewall coupling reduction technique (Fig. 3.12), while also blocking the high voltage by capacitors. The topology drives all the flying capacitors by one of the two pairs of bottom power switches (S_{1a} and S_{1b}, or S_{2a} and S_{2b}), as illustrated in blue in Fig. 3.12. This divides the flying capacitors into two distinct groups, each with its own synchronized voltage swings. By placing the capacitors of a group together in the layout, their mutual sidewall coupling is eliminated.

3.4.3 Toward High Power Density

The previous subsections introduced two topology techniques that achieve a good trade-off between efficiency and power density when operating at high input voltages. Within this trade-off, this work targets an exceptionally high power density, even if it comes at the expense of reduced efficiency. This pursuit of extreme power density introduces an additional challenge, as the maximum achievable frequency is inherently limited within high-voltage technologies, as earlier elaborated in Sect. 3.3.3.

To enhance power density, parameters that impact this metric are explored. The power density (P_{dens}) is derived in function of these parameters, commencing with the definition of P_{dens}:

$$P_{dens} = \frac{P_{out}}{A}, \tag{3.16}$$

$$= \frac{V_{out} \cdot I_{out}}{A}. \quad (3.17)$$

The output current in (3.17) can be calculated from the equivalent SCC schematic in Fig. 2.2:

$$I_{out} = \frac{\frac{V_{in}}{N} - V_{out}}{R_{out}}. \quad (3.18)$$

Subsequently, the output impedance in (3.18) is approximated when assuming operation in SSL. This condition occurs when the power switches are sized sufficiently large to maximize the power density, rendering the FSL output impedance negligible (2.17). Substituting the output impedance with the SSL output impedance from (2.14) yields the following result:

$$P_{dens} \propto f_{sw} \cdot \frac{C_{fly,tot}}{A} \cdot \left(\frac{V_{in}}{N} - V_{out} \right), \quad (3.19)$$

with $C_{fly,tot}$ representing the total equivalent flying capacitance.

Equation (3.19) reveals three ways for augmenting power density, given that the input and output voltages are fixed by the specifications. These are increasing the switching frequency, which is the conventional approach (see Sect. 2.2.4.2), elevating the flying capacitance while maintaining a constant area, and decreasing the topology conversion step. Nevertheless, further increasing the switching frequency proves unfeasible as the converter is already operating at its maximum switching frequency (f_{max}) (black graph in Fig. 3.13). The low maximum frequency stems from the large technology node required for high-voltage operation, as previously discussed in Sect. 3.3.3. Alternatively, implementing a larger flying capacitance would increase the area for a given capacitor type, eventually not gaining power density.

The third approach, opting for a lower topology conversion step, emerges as a feasible approach to boost power density, as illustrated in Fig. 3.13. In the equivalent model of Fig. 2.2, the $N : 1$ transformer now outputs a higher voltage, while a larger voltage drop across the output impedance maintains an identical V_{in}-to-V_{out} conversion. This increased voltage drop corresponds to an additional voltage differ-

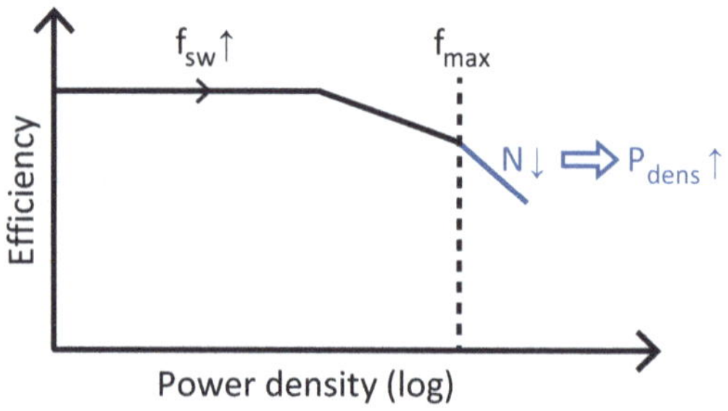

Fig. 3.13 Efficiency versus power density for a fully integrated SCC

ence between charge-sharing flying capacitors, facilitating more charge transferred per cycle. As a consequence, the output power and, thus, the power density are enhanced. Note, however, that this approach also leads to more SSL losses, leading to a suboptimal loss distribution, and higher-rated power switches. Nevertheless, this work chooses this last option in line with the target specification. This corresponds to realizing a 33:1 conversion by implementing a 24:1 topology.

3.5 Implementation

This section presents the implementation of the optimal high-voltage topology, which is shown in Fig. 3.14. It begins with the sidewall coupling reduction technique. The subsequent part delves into the custom design of both the flying capacitors and drivers for high voltage, ensuring minimal area overhead and a sufficient overdrive voltage, respectively. Following that, Sect. 3.5.4 introduces a technique to reduce the long latency of clocked hysteretic controllers. Lastly, the start-up details of the converter are discussed.

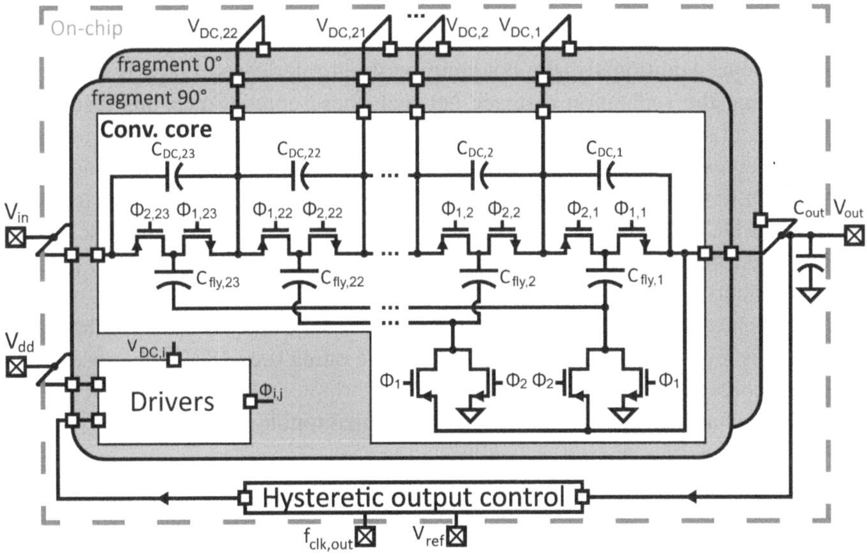

Fig. 3.14 System implementation, showing the transistor-level implementation of the converter cores

3.5.1 Sidewall Coupling Reduction

Even though physically separating flying capacitors with different swings decreases their mutual coupling (cf. Sect. 3.4.2), it also needs longer interconnections. The associated parasitic capacitances (e.g., to the substrate) introduce additional losses as they swing along with the flying capacitor, as illustrated in Fig. 3.15a. To counter these losses, the lengthy interconnections are made DC. A DC node is generated [8] by splitting the interconnecting power switch into two switches, creating a DC node in between, as illustrated in Fig. 3.15b. Although this DC node exhibits low impedance through its adjacent power switches and flying capacitors in phase 2, it shifts to high impedance during phase 1 and the nonoverlapping interval between the two phases. To achieve a consistent low-impedance path, these DC nodes are connected to the corresponding DC nodes of the interleaved converter cores, as shown in Fig. 3.15c. Specifically, the 180°-core establishes a low-impedance path during phase 1 of the 0°-core. Concurrently, the 90°- and 270°-cores also provide low impedance during the nonoverlapping intervals between the two phases of the 0°-core.

Another challenge is the coupling between different groups, as the sidewall coupling reduction technique only eliminates the coupling losses within a group. To reduce coupling within the same converter core, the flying capacitors of these different groups are spaced apart by positioning the power switches in between them. Additionally, the coupling between interleaved cores is decreased by increasing the separation distance between these cores. Since this distancing leads to an area overhead, the converter is only interleaved once. The converter comprises cores at 0° and 90°, while omitting the antiphase cores at 180° and 270°. To maintain low impedance during antiphase, each DC node is interconnected with its adjacent DC nodes via low-voltage DC capacitors ($C_{DC,i}$) of 400 fF each [15], as illustrated in Fig. 3.16. These DC capacitors provide a low-impedance path to these neighboring DC nodes, which are low impedance during antiphase, as their power switches are controlled in the opposite phase. The area overhead of these DC capacitors is minimal because of their low voltage rating (i.e., 18 V). The absence of antiphase cores does not affect the output-voltage ripple, as both phases still deliver an identical charge to the output in the implemented topology.

3.5.2 High-Voltage Flying Capacitors

Figure 3.17 shows the flying capacitor implementation (3.8 pF each). TP and BP represent the top and bottom plates of the capacitor in the SCC topology, respectively. Beyond using the standard metal plates, it uses a diffusion layer (n+) in the n-well as an additional capacitor plate, thereby enhancing the capacitance density.

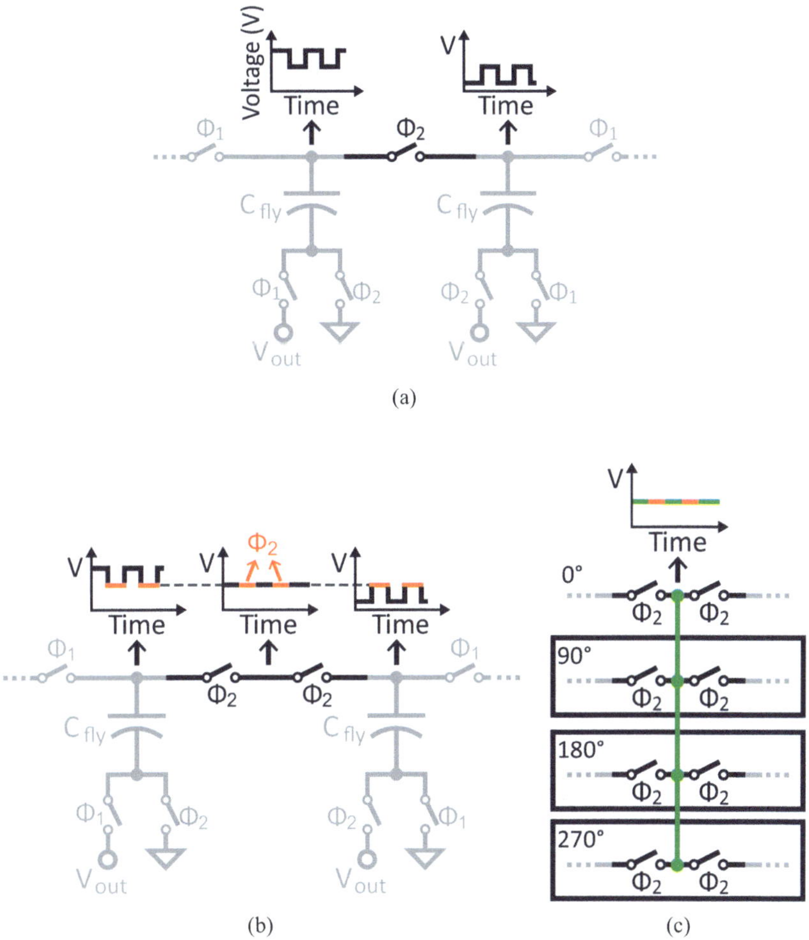

Fig. 3.15 DC nodes generation. (**a**) Starting from the Dickson-star topology, (**b**) splitting interconnecting power switches, and (**c**) connecting corresponding DC nodes of interleaved converter cores

Conventionally, the diffusion plate couples to the p-substrate with only the BOX capacitor (C_{BOX}) in between, leading to a large parasitic capacitance. The incorporation of a handle-wafer diode reduces the parasitic coupling of a flying capacitor to the substrate [59]. These diodes are available in high-voltage SOI technologies to realize high-voltage diodes and switches (cf. Sect. 3.3.1). Figure 3.17 demonstrates how inserting a reverse-biased diode beneath the flying capacitor places its junction capacitor (C_j) in series with the preexisting C_{BOX}. This leads to a reduction in the overall substrate capacitance (C_{sub}).

The diode is generally biased with the maximum reverse voltage (V_{bias}) to reduce substrate capacitance, as this maximizes its depletion width. For this purpose, a bias

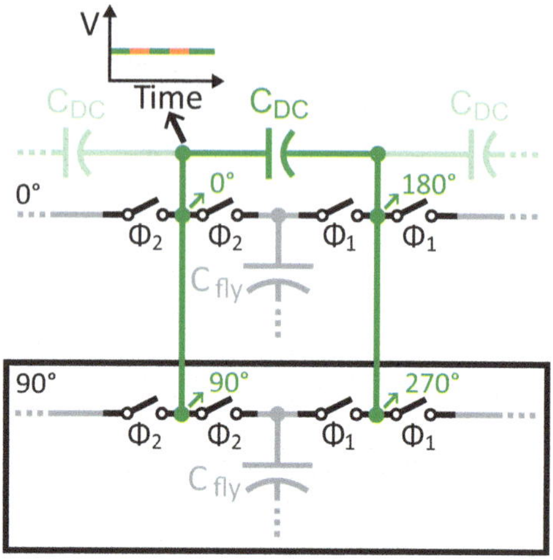

Fig. 3.16 Low-voltage DC capacitors connecting neighboring DC nodes to maintain low impedance when eliminating antiphase cores

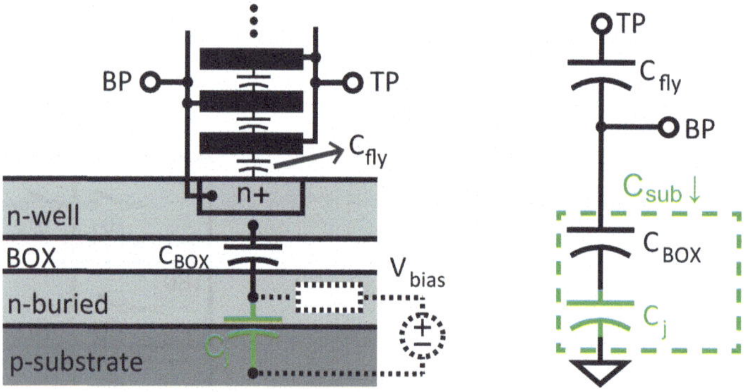

Fig. 3.17 (**a**) Reducing the substrate coupling of the flying capacitor by placing a handle-wafer diode below the BOX, (**b**) with the AC equivalent schematic

voltage source is directly connected to the n-buried layer in Fig. 3.17a. Despite the direct connection, C_j is not shorted, as the path through the n-buried layers is highly resistive. This high resistivity stems from the complete depletion of the n-buried layer, in which the depletion layer touches the buried oxide. This full depletion occurs at a relatively low voltage, such as 40 V in [59].

While a handle-wafer diode reduces the substrate coupling, the area overhead is enormous. The high-voltage electric field in this diode requires wide spacings with DTI structures for termination (cf. Sect. 3.3.1). These spacings are around every flying capacitor, of which many are inherently needed in high step-down SCCs [37]. This results in a quadrupling of the required area for the same flying capacitance used in this design.

To minimize this termination overhead, this work proposes three improvements (A, B, and C in Fig. 3.18). First (A), the n-buried layer is biased at its minimum

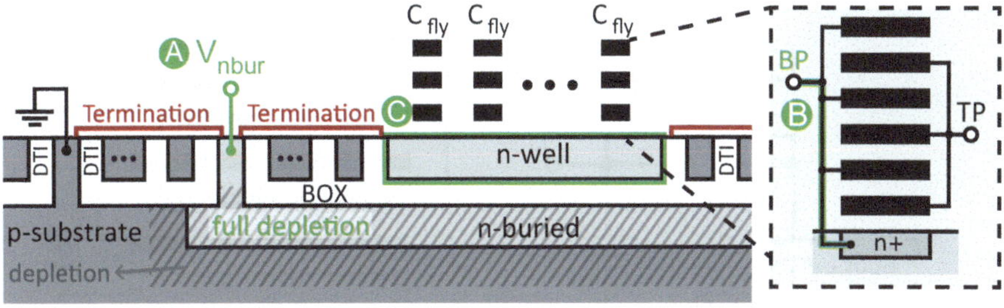

Fig. 3.18 Implementation of the high-voltage flying capacitors (C_{fly}), with improvements A, B, and C that limit the area overhead.

voltage (V_{nbur}) where it reaches fully depleted, rather than applying the highest voltage available as in [59]. Raising this bias voltage beyond full depletion only insignificantly decreases the substrate capacitance [59], while demanding significantly more area for field termination. The bias voltage is generated externally for testing flexibility but can be biased by an intrinsically available DC node on-chip ($V_{DC,i}$ in Fig. 3.14).

Second (B), the flying capacitor plate closest to the substrate (i.e., diffusion plate) is connected to the terminal with the lowest voltage (BP). This circumvents the need for extensive field termination structures. This is in contrast to conventional practice where this plate is connected to (TP) to enable charge intrinsic charge recycling [10, 42]. The gain of the latter would be marginal because of the reduced substrate capacitance by the handle-wafer diode.

Third (C), as an additional benefit of the sidewall reduction technique, the bottom-plate terminals of adjacent flying capacitors are shorted together. A shared diffusion plate enables a single handle-wafer diode beneath all flying capacitors within the same group, while avoiding extensive field termination between the capacitor units. Combining the three techniques discussed leads to a reduction in area overhead by a factor of 2.7.

3.5.3 Drivers

The bottom nMOS power transistors in Fig. 3.14 are driven directly by an inverter chain, as their source voltage aligns with the ground. In contrast, all other power transistors have a non-ground source voltage, requiring level shifting. Figure 3.19a depicts the capacitive bootstrap circuit that drives the top nMOS power switches, with V_S representing the source of these switches and $\Phi_{i,j}$ the gate of the jth top power switch controlled by phase i. The driver comprises a right bootstrap capacitance ($C_{boot,R}$) of 170 fF and a left one ($C_{boot,L}$) of 40 fF. The circuit driving the pMOS power switches is identical, except that transistors M_1 and M_2 become pMOS. The applicability of this simple bootstrap driver to all power switches

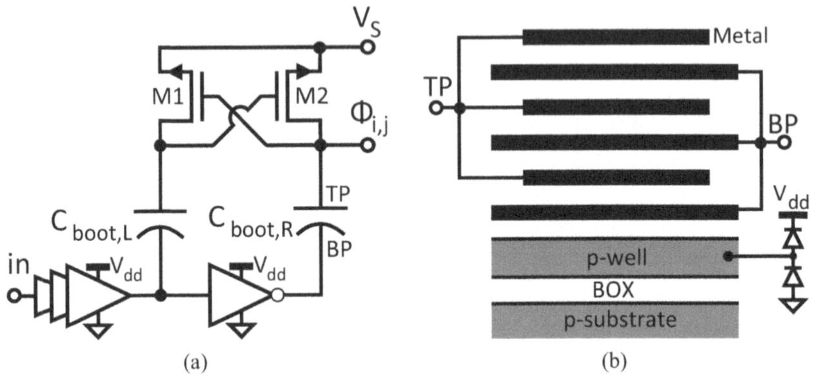

Fig. 3.19 (**a**) Driver implementation for top nMOS power switches. (**b**) Side view of the bootstrap capacitors

is because of their DC source voltages, such as the generated DC nodes ($V_{DC,i}$ in Fig. 3.14) for the top power switches. In case no DC nodes would have been generated, their source voltages would be periodically swinging with an amplitude of $\sim$12 V ($> V_{dd}$) at high voltage (cf. Fig. 3.15a). This scenario would necessitate a lossy level-shifter, as well as an area-expensive and power-hungry charge pump like the self-boost variant in [35].

The bootstrap driver operates effectively at low voltage for which high-quality capacitors are available. At high voltage, however, the bootstrap capacitors encounter increased parasitic coupling to the substrate and to neighboring bootstrap capacitors as they block a higher voltage. Unlike the flying capacitors, they do not have a handle-wafer diode below the buried oxide (BOX) to reduce substrate coupling. Incorporating such a diode would increase the area of the drivers immensely to terminate the high-voltage fields. Consequently, the large parasitic capacitances at high-voltage result in an insufficient overdrive voltage (V_{ov}) of the power switch, which is described as follows:

$$V_{ov} = \frac{C_{boot,R}}{C_{boot,R} + C_{par,HV}} V_{dd} - V_{th}. \tag{3.20}$$

Herein, V_{dd} is the supply voltage, V_{th} denotes the threshold voltage of the driven power switch, and $C_{par,HV}$ represents all parasitic capacitances in the high-voltage domain on node $\phi_{i,j}$ in Fig. 3.19a. The latter includes the top-plate (TP) parasitic capacitance of the bootstrap capacitor and the gate capacitance of the power switch. Notably, the bottom-plate (BP) capacitance of the bootstrap capacitor does not compromise the overdrive voltage.

To increase the overdrive voltage, the parasitic top-plate capacitance is reduced. Figure 3.19b illustrates that the bootstrap capacitors are customly designed to shift their parasitic coupling from the top plate to the bottom plate, because the BP capacitance does not lower V_{ov} (3.20). To achieve this, the bottom plate of the capacitors shields the top plate from the substrate. At the sides, the bottom plates extend outward such that most sidewall coupling occurs at the bottom plate rather than at the top plate.

In addition, $C_{par,HV}$ is further minimized by directly driving the power switch gate by the buffers in the low-voltage domain through $C_{boot,R}$, rather than relying on buffers in the high-voltage domain as in [15]. In the latter study, the level-shifter capacitor ($\approx C_{boot}$) also charges the parasitic capacitances of the buffers in the high-voltage domain, corresponding to a higher $C_{par,HV}$. In another approach [35], a dedicated charge pump instead of the level-shifter capacitor supplies this charge, but this necessitates a complex and large implementation because the source voltage never returns to the ground.

Besides enabling high-voltage operation, the driver is also customized to diminish the driving losses. This is accomplished by reducing substrate coupling through the high-impedance biasing of the p-well in Fig. 3.19b, similar to the method described in Sect. 3.5.2 [10, 59]. In AC, the BOX capacitor is introduced in series with the BP-to-p-well capacitor, reducing substrate coupling. In DC, the p-well is biased by two diodes. While a conventional high-impedance approach [10, 59] could suffice, using two diodes offers an additional low-impedance path outside the rails for additional protection against breakdown.

3.5.4 Control

The converter regulates the output voltage by employing a lower-bound clocked hysteretic controller based on [64], as discussed earlier in Sect. 2.3.1. This control method allows to instantaneously adjust the effective switching frequency, which is required for a fast load-step response. As shown in Fig. 3.20a, the controller monitors every clock cycle ($f_{clk,out}$) whether the output voltage (V_{out}) is below the constant reference voltage (V_{ref}). If so, it triggers the converter to deliver a charge packet to the output.

However, the clocked operation introduces a considerable latency (Fig. 3.20a). When V_{out} drops below V_{ref} right after a rising clock edge, the controller only starts responding after a full clock period. During this delay, particularly under heavy load, the voltage continues to drop significantly. Afterward, the converter delivers a charge packet to the output to restore V_{out} above V_{ref}. However, under heavy load, a significant part of this charge goes to the load instead of charging the output capacitor. As a result, the controller cannot raise V_{out} above V_{ref} before the next rising clock edge, triggering a second, undesired charge event. This increases the ripple under heavy load, dominating the overall ripple, and introduces a droop at a light-to-heavy load step.

To address this high ripple, a prior work [79] splits its SCC into multiple fragments (identical parallel SCC units), of which some are operating at the maximum frequency. This latter part is constantly delivering power to the output,

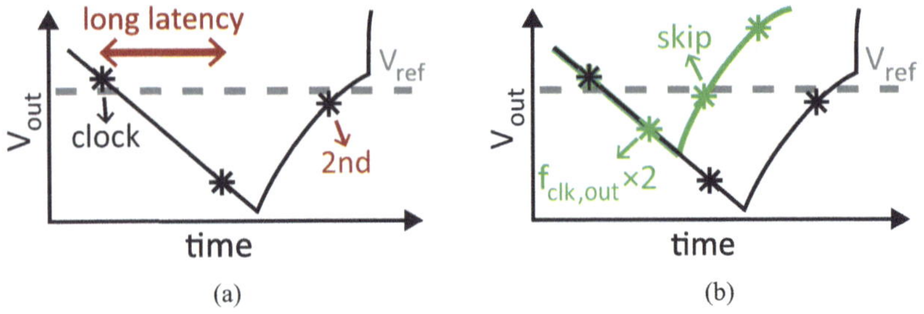

Fig. 3.20 (**a**) Conventional clocked hysteretic control leads to a second undesired charge event. (**b**) Double clocking and skipping second-in-a-row charge events avoid this undesired charge event

mitigating the voltage droop. As a result, the controller can recover the output voltage in time. A secondary control loop dynamically adjusts the number of fragments operating at maximum frequency. However, this second control loop evaluates the load condition based on multiple prior decisions of the conventional hysteretic controller, resulting in a slow process. While this technique reduces ripple in steady state, it cannot handle rapid load transients. Furthermore, this method introduces additional losses, as some fragments are switched at the maximum frequency even when the load current does not require it.

Alternatively, this work clocks the comparator at double the frequency, as shown in Fig. 3.20b. This halves the worst-case latency, reducing the voltage drop before the controller starts reacting. However, this increased clock frequency also accelerates the occurrence of the subsequent clock edge, leaving only half the original time for the control loop to restore the output voltage. Consequently, an unintended second charge event still occurs. As a solution, besides the doubled clock frequency, every second-in-a-row charge event is skipped. This approach maintains the halved latency while providing the controller the original time to restore the output voltage. This technique is extendable to higher comparator frequencies with additional skipped charge events. Nevertheless, in this particular design, this extension is unnecessary since the heavy-load ripple is not dominating anymore after doubling the comparator frequency.

Figure 3.21 depicts the relevant part of the output control implementation, specifically focusing on the introduced skipping circuit. The first flip-flop stores whether a charge event occurred in the previous clock cycle. If there was no charge event, the nand-gate passes the output of the comparator, and vice versa. Since the output of the implemented StrongARM comparator [56] is only valid for half the clock period, a first flip-flop stores its value at the falling clock edge. To prevent glitches, a direct connection between the output of the first flip-flop and the nand-gate is avoided. Such a connection would result in the first flip-flop's output reaching the nand-gate before the comparator had fully reset to 0 at the negative clock edge, potentially introducing glitches. Instead, a second flip-flop serves as a delay element, mitigating this issue.

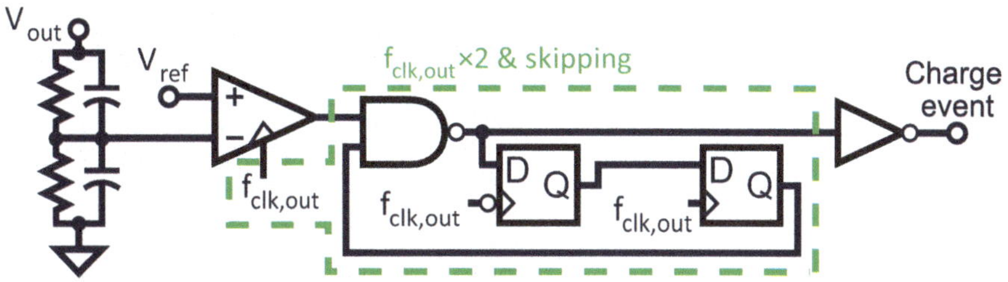

Fig. 3.21 Implementation of the improved control circuit

3.5.5 *Start-up*

To start up the converter, the following procedure is executed. First, the control voltages are raised. Although these control voltages are supplied externally for testing flexibility, they can be generated on-chip during both start-up and steady-state operation, as will be detailed in Sect. 3.6.1. Second, the drivers are clocked to set their latches to the correct state. While this step could introduce current conduction through the bulk diodes, it does not lead to latch-up. This is because the latches consist solely of nMOS or pMOS transistors, forming no parasitic thyristor. Additionally, these complementary transistor types are isolated by oxide because of the SOI technology. The third step ramps the DC input voltage from 0 V to 400 V at 0.1 V/µs. Since the controller clocks at 200 MHz, the flying capacitors inherently pre-charge themselves, eliminating the need for a dedicated initialization circuit.

3.6 Chip Measurements

A prototype has been implemented using all described concepts. The design has been fabricated in a 180 nm high-voltage CMOS SOI technology. While fully integrated, the prototype converts a 400 V DC input into a stable 12 V DC output and delivers a peak output power of 220 mW. The micrograph in Fig. 3.22 shows an active area of 1.85 mm^2. This includes the integrated output decoupling capacitor (C_{out}, 128 pF), which consumes 9% of the active area.

3.6.1 *Measurement Setup*

Figure 3.23 overviews the measurement setup. The input voltage is provided by a Keysight 6812B voltage source. Due to the limited measurement accuracy of this input source, the input power is measured by two Keysight 34401A multimeters. One of the multimeters measures the voltage across a shunt resistor (R_{shunt}) to

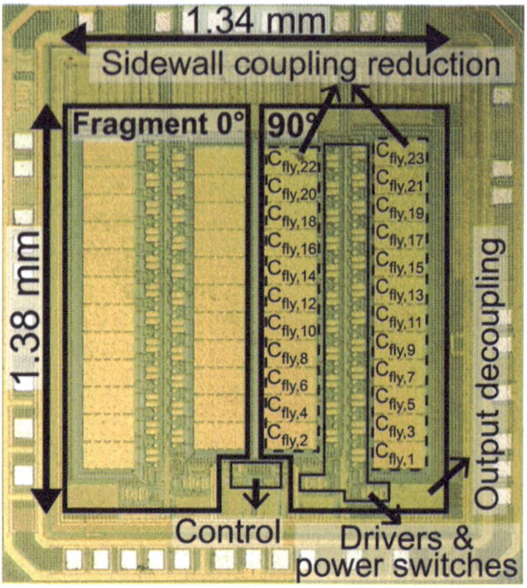

Fig. 3.22 Die micrograph of the fully integrated DC-DC converter, with 1.85 mm^2 active area

determine the input current. The other multimeter measures the input voltage before the shunt resistor and compensates for the voltage drop across it (the latter is measured by the current-measuring multimeter). An alternative placement of the voltage multimeter after the shunt resistor is not suitable. This would introduce an additional current through the shunt resistor due to its input resistance to the common ground for the voltage measurement. This 1 MΩ input resistance would incorrectly elevate the measured input power by 160 mW because of the high input voltage of 400 V.

The 5 V driver supply (V_{dd1}) and the 1.8 V control supply (V_{dd2}) are powered externally by a low-dropout regulator (LDO) to enhance testing flexibility. Their power is measured by a shunt resistor, a current-sense amplifier, and an analog-to-digital converter (ADC). It is worth noting, however, that these two supplies could be generated on-chip with limited additional area overhead and power loss. At start-up, shunt regulators in parallel can temporarily generate these supply voltages from the input voltage. Once the converter is operational in steady state, the shunt regulators are deactivated to minimize losses. In this case, the supply rails can be efficiently generated from the 12 V DC output of the main SCC by using an auxiliary low-voltage SCC. This efficient generation, combined with the low power requirements of the supplies, would result in a negligible additional power loss. Specifically, the power demand for the driver supply amounts to only 5% of the output power, as it scales with the switching frequency, while the control supply consumes a mere 206 μW.

The reference voltage for output control (V_{ref}) and the n-buried bias voltage (V_{nbur}) are also generated externally for measurement flexibility. The voltage of V_{ref} is provided by a digital-to-analog converter (DAC). The associated losses are not measured as they are negligible, given that V_{ref} is a low-voltage bias for the output control comparator. Furthermore, a Keithley 2612A source meter generates

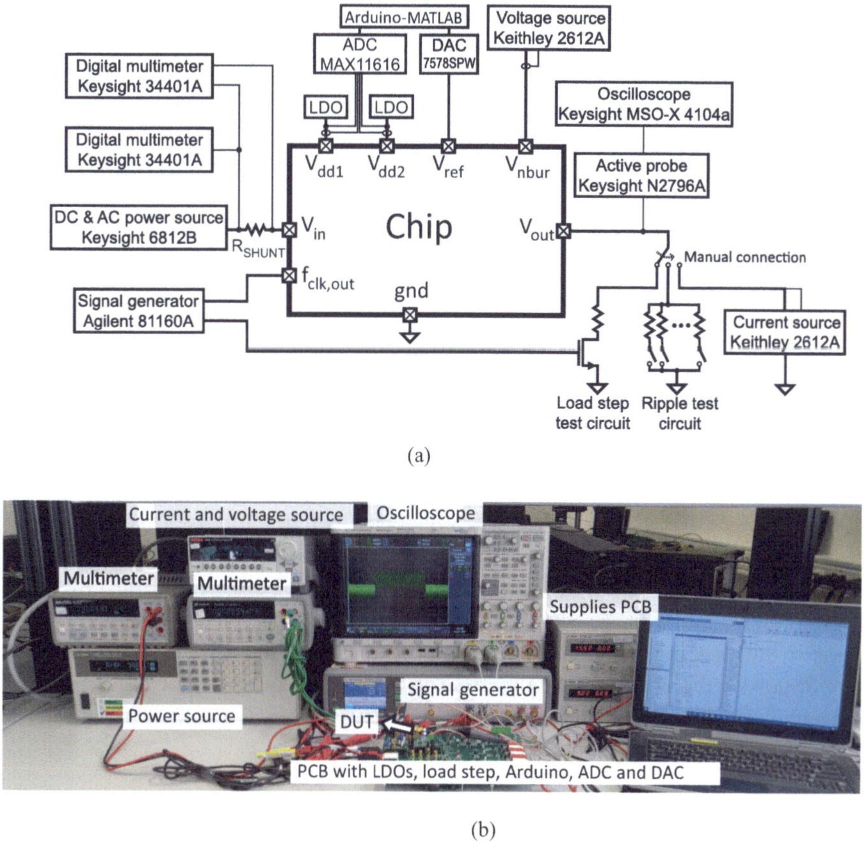

Fig. 3.23 Measurement setup. (**a**) Schematic overview. (**b**) Picture of the test equipment in the lab

V_{nbur} and measures its power consumption. As mentioned before, the DC voltage for V_{nbur} could be provided on-chip, as it is readily available among the generated DC nodes. The node V_{nbur} requires minimal current because of its high impedance (cf. Sect. 3.5.2).

Three distinct load circuits are employed to address the specific measurement requirement. First, a Keithley 2450 source meter functions as a current source for the efficiency measurements. It conducts an automatic sweep across the load current range while continuously monitoring the output voltage. Despite its convenience for automatic efficiency measurements, this method is unsuitable for load-step and ripple measurements due to the extended connections. These long connections introduce additional output capacitance, which would incorrectly enhance the measured ripple. In response to this limitation, the subsequent load circuits are designed on a printed circuit board (PCB) close to the chip output, ensuring that the impact of the additional capacitance becomes negligible. The second load circuit serves for load-step measurements, employing a MOSFET in series with a load resistor. This transistor is driven by an Agilent 81160A signal generator, enabling rapid activation and deactivation of the load resistor. Concurrently, the output voltage is captured by a Keysight MSO-X 4104a oscilloscope paired with a Keysight N2796A active probe. In the third load circuit, a resistor bank serves as the load to evaluate the output-voltage ripple under varying output power.

Finally, a MATLAB environment controls both the measurement PCB via an Arduino Uno and all the measurement equipment. It sets all the voltages such as the input voltage to enable automated measurements. It also collects the data of the power measurements and calculates the efficiency.

3.6.2 *Efficiency*

Figure 3.24 shows the measured full-system efficiency versus its output power for a 400 V-to-12 V DC-DC conversion. The efficiency includes all losses except those to generate the external reference bias and the external clock. The converter attains an efficiency of 63.6% at a power density of 119 mW/mm^2. Furthermore, the efficiency curve remains flat (>60%) across a wide power range between 4 mW and 220 mW, because the switching frequency scales proportionally to the output power. At lower output power, the efficiency begins to drop because frequency-independent losses start to dominate (e.g., comparator losses). This leads to a quiescent power of 266 μW.

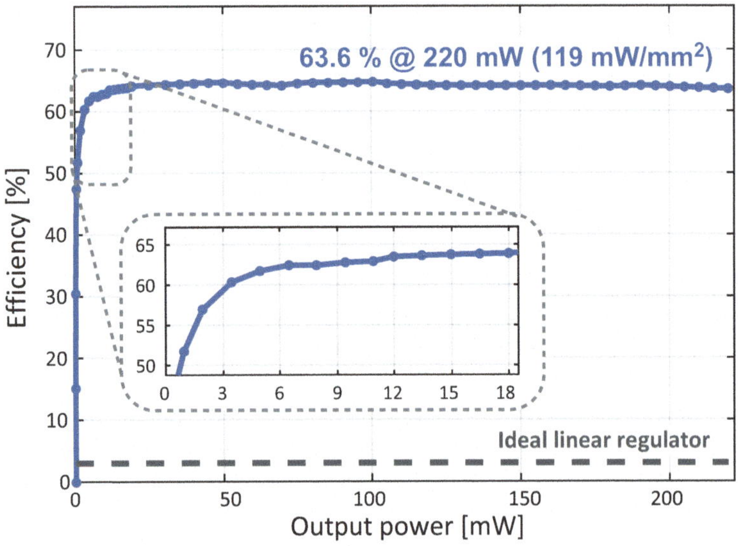

Fig. 3.24 Measured full-system efficiency versus output power for a 400 V-to-12 V DC-DC conversion (blue). The gray dotted line shows the ideal linear regulator efficiency for the same conversion step

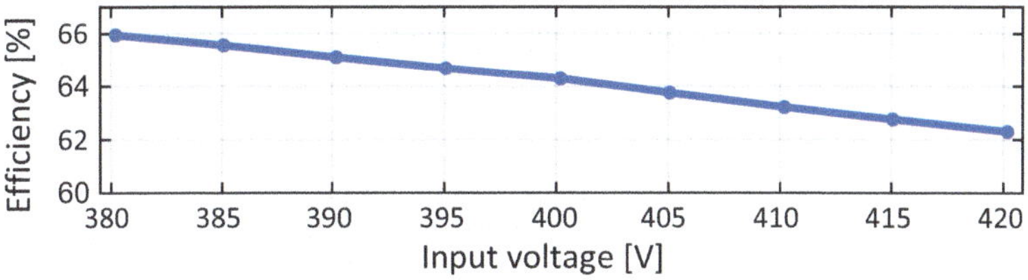

Fig. 3.25 Measured full-system efficiency versus input voltage for an output voltage of 12 V and an output power of 185 mW

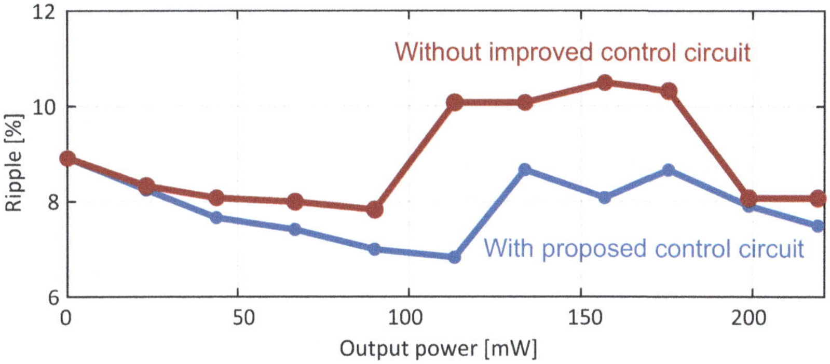

Fig. 3.26 Measured ripple versus output power with improved control technique (blue) and without (red)

Figure 3.25 shows the converter ability to operate over a DC input range from 380 V to 420 V for an output power of 185 mW, while regulating the output voltage at 12 V DC. Reducing the input voltage while keeping the output voltage constant necessitates a smaller voltage drop across the output impedance. Consequently, the output power must decrease, given that the converter is already operating at its maximum frequency (3.19).

3.6.3 Transients

Figure 3.26 depicts the measured ripple versus the output power with the improved control method (blue) and without (red). As previously discussed, this method lowers the ripple at a heavy load, leading to an overall ripple reduction of 14.5%. At peak load, the controller reaches maximum switching frequency for both options, resulting in nearly identical ripple levels.

Figure 3.27a demonstrates the output response of an applied load step going from maximum load to zero load and back with a transition time (Δt) of 25 ns. The output response shows no overshoot or droop, whereas disabling the improved

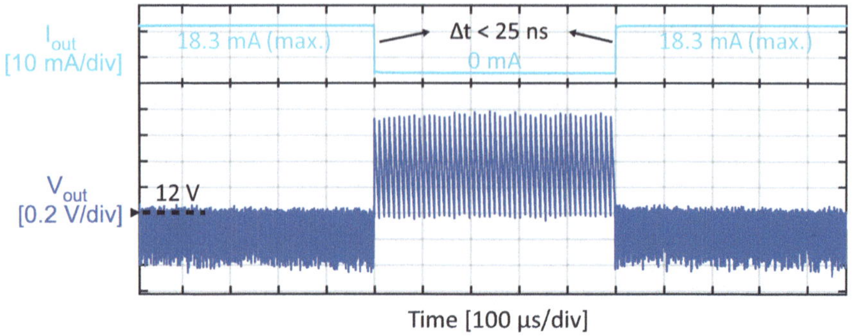

(a)

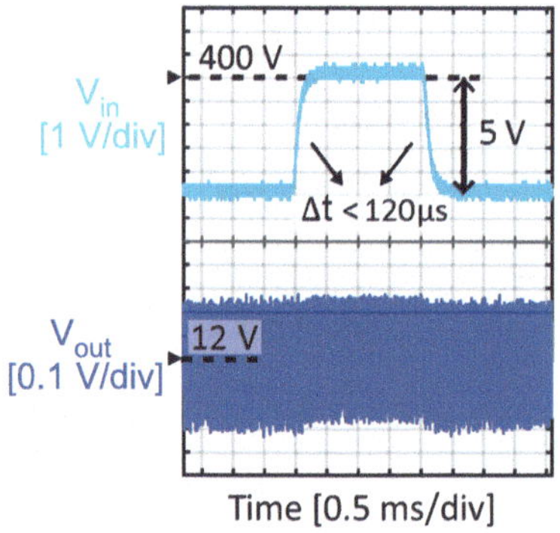

(b)

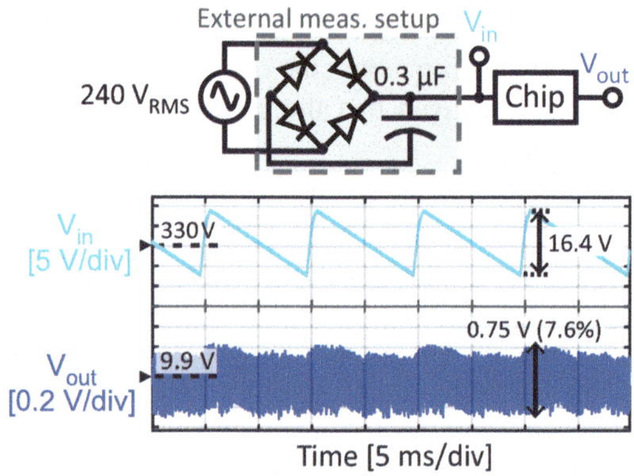

(c)

Fig. 3.27 Measured worst-case transients. (**a**) Load-step response with DC input voltage V_{in} = 400 V and reference voltage V_{ref} = 12 V. The converter switches from maximum load to zero load and back within 25 ns. (**b**) Line step at maximum load. (**c**) Line regulation for a rectified mains input at 126 mW

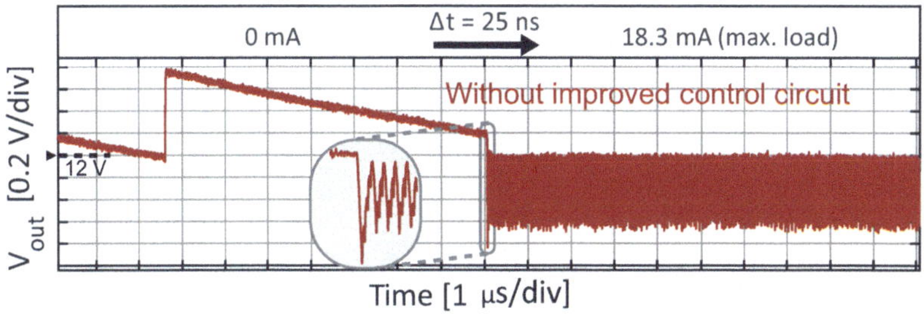

Fig. 3.28 Measured load-step response without the improved control technique

control technique leads to a droop (Fig. 3.28). The line regulation under a line step at maximum load also displays no overshoot or droop (Fig. 3.27b). Lastly, Fig. 3.27c shows the compatibility with a rectified European mains input of 240 V_{RMS}. The measurement setup incorporates an external buffer capacitor of 0.3 μF and external rectifier diodes.

3.6.4 Comparison

Table 3.1 compares this work with state-of-the-art and commercial DC-DC converters with high input voltages, along with a linear regulator. First, the linear regulator [44] reaches an efficiency (η) of only 2.1% and requires an external capacitor. Second, switching DC-DC converters with a similar input voltage [22, 25, 58, 59] requires at least two external components [59], including a large inductor. This results in a converter that is expensive, bulky, and less reliable. Third, state-of-the-art fully integrated converters [15, 35] only attain input voltages up to 42 V. In contrast, this work pushes this input voltage to 400 V while advancing the power density by 270×. Furthermore, it realizes the highest conversion step and the highest efficiency at the maximum input voltage $V_{in,max}$.

3.7 Conclusion

This chapter developed the first fully integrated 400 V DC-DC converter. Previous fully integrated converters could only handle input voltages up to 42 V and exhibited poor performance due to inferior component quality and increased parasitics at high input voltages. To overcome these limitations, this work maximized the dielectric energy storage at high voltage, substantially minimized the sidewall coupling losses of flying capacitors, and optimized the conversion step. The custom design of the high-voltage capacitors and drivers further enhanced the performance. Additionally,

Table 3.1 Comparison with state-of-the-art and commercial DC-DC converters with high input voltages

	Fully integrated			Not fully integrated					
	This work	JSSC20	JESTPE18	JSSC21	LR645	CICC22	ISSCC20	MuxCap	JSSC16
		[15]	[35]	[59]	[44]	[58]	[25]	[22]	[94]
Technology [nm]	180, SOI	350	350	180, SOI	/	180, SOI	GaN	180, SOI	500
Tech. V_{max} [V]	400	42	700	400	/	400	650	400	120
Topology	SCC	SCC	SCC	Resonant	Linear regulator	Flyback	Buck	SCC	Multi-level
Input voltage [V]	**400**	7.5–**42**	**17**	100–325	15–450	20–300	60–400	<400	12–100
Output voltage [V]	12	3	3.3	3.3–10	10	5	30–60	5	10
Max. conversion	33:1	14:1	5:1	98:1	45:1	60:1	13:1	80:1	10:1
# external comp.	**0**	0	0	**2**	1	**7**	**7**	**12**[a]	**3**
P_{max} [mW]	220	2.1	3	500	30	500	29000	250	5000
Power density [mW/mm^2]	**119**	**0.4**	**0.44**[a]	3.2[b]	1.8[a]	0.94[b]	/	/	/
η @ $V_{in,max}$ [%]	63.6	50.3	47.4	61.9	**2.1**	60.4	88.8[a]	>90	82.5
Max. ripple [%]	9.2	5	6	6	0	3.7	1	/	5

[a] Estimated
[b] Estimated with PCB area

an improved control circuit resulted in an excellent step response and an overall ripple reduction.

In comparison to the state of the art, this work eliminates external components. It improves the input voltage of fully integrated step-down converters by 9× and the power density by 270×. Alongside achieving the largest conversion step, it also attains the highest efficiency in class.

Although this chapter has successfully met the predefined objectives for a DC input voltage, it still requires external components for compatibility with the AC mains. The following chapters extend this converter to accommodate an AC mains input.

Chapter 4
Gearboxing a Fully Integrated High-Voltage DC-DC Converter

The previous chapter developed a fully integrated high-voltage switched-capacitor converter (SCC) with a DC input voltage. This chapter, along with the subsequent one, targets an AC input voltage, thereby unlocking additional applications powered by the ubiquitous AC mains. While facilitating an AC input voltage, the other specifications of Sect. 3.1 such as complete integration and high power density persist.

In prior fully integrated AC-DC converters, the AC-DC stage not only rectifies the AC input but also performs the down-conversion. They employ a capacitive divider, leveraging the suitability of capacitors for integration. However, the combination of the low AC mains frequency (50 or 60 Hz) and the low capacitance density on-chip results in a high input impedance. This high impedance prevents the flow of large currents, ultimately restricting the power density to values $\leq 1.6\,\mu W/mm^2$ [14].

To overcome the dependence on the mains frequency, this chapter investigates a fully integrated DC-DC stage following the AC-DC stage to perform the down-conversion. The ability to independently set the switching frequency of the DC-DC converter, irrespective of the AC mains, enables higher power densities. This approach requires a DC-DC converter handling the slowly varying rectified AC input, spanning from 0 to 325 V (assuming the European mains). The DC-DC converter introduced in the previous chapter is incompatible with this input range due to its fixed ideal voltage conversion ratio (iVCR). In response, this chapter extends the high-voltage SCC from the preceding chapter into a multi-ratio (or gearbox) converter, as a first step toward AC-DC conversion.

The chapter starts by explaining why a wide-input-range DC-DC converter can handle a rectified AC input voltage. The following part shows that state-of-the-art fully integrated DC-DC converters cannot handle the input voltages required for the AC-DC application and delves into the specific challenges. To address these challenges, a high-voltage multi-ratio DC-DC topology is proposed. Subsequently, the implementation details are presented, covering the high-voltage driver design

© The Author(s), under exclusive license to Springer Nature Switzerland AG 2025
T. Van Daele, F. Tavernier, *Fully Integrated High-Voltage DC-DC and AC-DC Conversion*, Analog Circuits and Signal Processing,
https://doi.org/10.1007/978-3-031-78519-1_4

and high-voltage input sensing. Finally, this chapter shows the measurement results obtained from the prototype and compares them with state-of-the-art solutions.

Parts of this chapter were presented at the Symposium on Very Large-Scale Integrated Circuits (VLSI) of 2023 in Kyoto, Japan [84], and have been published in the Journal of Solid-State Circuits (JSSC) in 2023 [86].

4.1 Why Wide-Input-Range DC-DC for AC-DC Conversion?

This chapter develops a switched-capacitor (SC) DC-DC converter tailored for AC-DC conversion. After the AC mains input is rectified, the SC DC-DC converter down-converts the rectified AC input voltage into a low DC output voltage. Despite the DC-DC name suggesting an exclusive acceptance of fixed DC input voltages, this SC DC-DC converter handles the rectified AC input voltage. This capability stems from the slow variation of the rectified input voltage at a frequency of approximately 100 Hz, as shown in the left part of Fig. 4.1. In comparison, the switching frequency of the SC DC-DC converter (e.g., 50 MHz in Chap. 3) is substantially faster. As a result, the rectified AC input is effectively perceived as a fixed DC input by the SC DC-DC converter, as demonstrated by the zoomed-in view in Fig. 4.1. Thus, the SC DC-DC converter is a feasible solution to down-convert the rectified AC input, similar to the inductive converter in [60].

The SC DC-DC converter needs a wide DC input voltage range to convert the entire range of rectified AC input voltages. Ideally, it realizes a 0–325 V input range, assuming the European mains. However, in practice, a buffer capacitor takes over around the zero-crossing of the AC input, as detailed in Chap. 5. Nevertheless, to minimize the buffer capacitance, the minimum input voltage should be low. This requires a wide input voltage range of up to 325 V for the SC DC-DC converter.

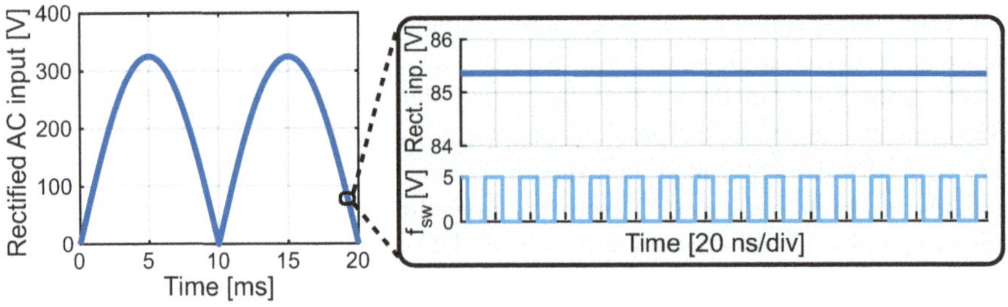

Fig. 4.1 Rectified AC input voltage, of which a zoom-in view is compared with a clock operating at a 50 MHz switching frequency (f_{sw})

4.2 State of the Art and This Work

The DC-DC converter demands a broad input range at high voltage while being fully integrated. However, prior works that achieved a wide and high input voltage range cannot meet the integration requirement. The underlying limitation aligns with the explanation previously given in Sect. 3.2 for high-voltage fully integrated DC-DC converters and, therefore, is not reiterated here.

On the other hand, existing fully integrated DC-DC converters cannot achieve a wide input voltage range at high voltage. Figure 4.2 visualizes their input voltage range, expressed by the iVCR range, versus their maximum input voltage. Previous works achieved either a wide input range [15] or a high input voltage (the converter in Chap. 3) but not both simultaneously. In particular, existing multi-ratio converters are limited to 42 V input, while the 400 V converter of Chap. 3 only supports a single conversion ratio.

Therefore, this chapter extends the high-voltage single-ratio SCC of the previous chapter into a multi-ratio converter. Each topology ratio handles the specific interval of rectified input voltages where it is most efficient. The division of the input voltage range into intervals is illustrated in Fig. 4.3. In comparison to existing fully integrated multi-ratio DC-DC converters, the proposed converter extends the supported input voltage by >7×.

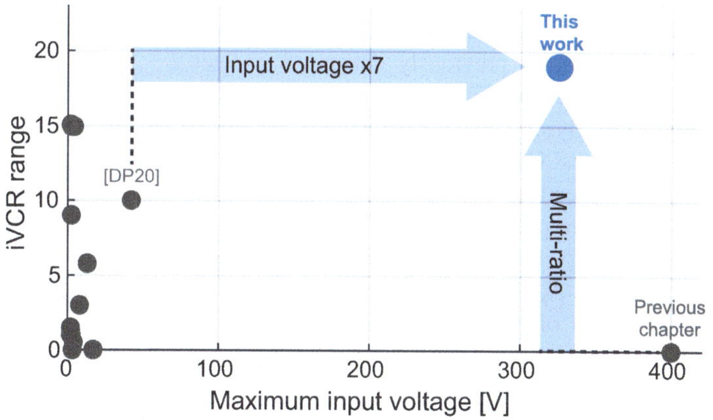

Fig. 4.2 iVCR range versus maximum input voltage for state-of-the-art fully integrated DC-DC converters, based on [69]

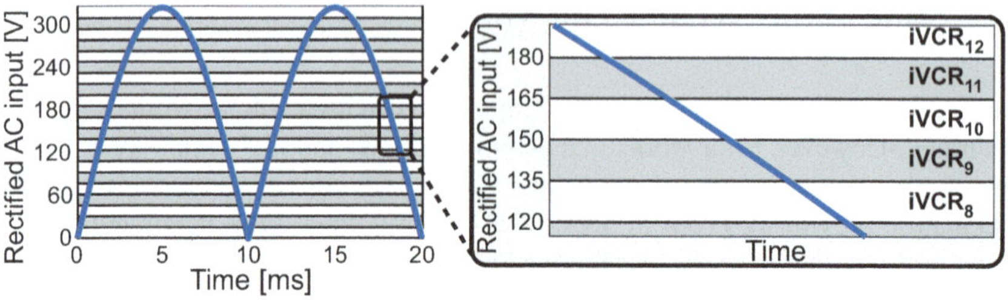

Fig. 4.3 The rectified AC input voltage is segmented into voltage intervals, each corresponding to a specific topology iVCR

4.3 Challenges

Widening the input voltage range of the single-ratio SCC of Chap. 3 introduces two critical topology challenges: an increased output impedance and an elevated output-voltage ripple. These two challenges are discussed in the following paragraphs.

4.3.1 Increased Output Impedance

Facilitating an extensive DC input voltage range in an SCC topology requires a wide range of iVCRs, as previously discussed in Sect. 2.3.2. These iVCRs are achieved through the reconfiguration of flying capacitors into various topologies. Reconnecting these flying capacitors mandates additional switches, referred to as configuration switches. However, the placement of these configuration switches in the conduction path leads to a notable increase in the FSL output impedance.

To illustrate this impact, consider the prior multi-ratio Dickson-star topology of [64]. This topology is designed to achieve a range of conversion ratios from $1/2$ to $1/N_{max}$, with N_{max} denoting the maximum ideal voltage conversion step. The FSL impedances of both the configuration switches (R_{config}) and the power switches (R_{power}) are assessed as a function of N_{max}, where higher values of N_{max} correspond to a wider input voltage range. The FSL impedances are calculated for a 2:1 conversion using (2.17), assuming an identical on-resistance R_{on} and a duty cycle of 50% for all switches. This yields the following expressions:

$$R_{power} = \frac{2N_{max} - 1}{2N_{max} - 2} \cdot R_{on} \simeq R_{on}, \tag{4.1}$$

$$R_{config} = \frac{N_{max} - 2}{2} \cdot R_{on} \simeq N_{max} \cdot \frac{R_{on}}{2}. \tag{4.2}$$

These formulas demonstrate that, as the input voltage range widens in multi-ratio topologies (N_{max} ↑), the impedance of the configuration switches becomes increasingly dominant. For instance, in the implemented design of this chapter, where $N_{max} = 21$, the FSL impedance of the configuration switches would be approximately 9× greater than the power switches, drastically degrading the converter efficiency.

4.3.2 *Elevated Ripple*

In addition to the increased output impedance, extending the SCC input range (DC) also introduces an elevated output-voltage ripple. SCCs achieve a wide DC input range by adjusting their iVCR and output impedance to accommodate varying input voltages (cf. Sect. 2.3). Specific combinations of iVCR and output impedance values can yield significantly higher ripple levels compared to other combinations. This phenomenon is demonstrated in Fig. 4.4, which illustrates the measured ripple versus input voltage of a prior Dickson-star topology [64], with conversion ratios ranging from 1/2 to 1/5. The 1/5 ratio exhibits a relatively low ripple, implying that fixed-step implementations like in Chap. 3 can have a low maximum ripple. However, when the input range is expanded, a significantly higher maximum ripple appears. Given that it is the maximum ripple, rather than the average, that determines the required output capacitance, this results in a substantial area increase.

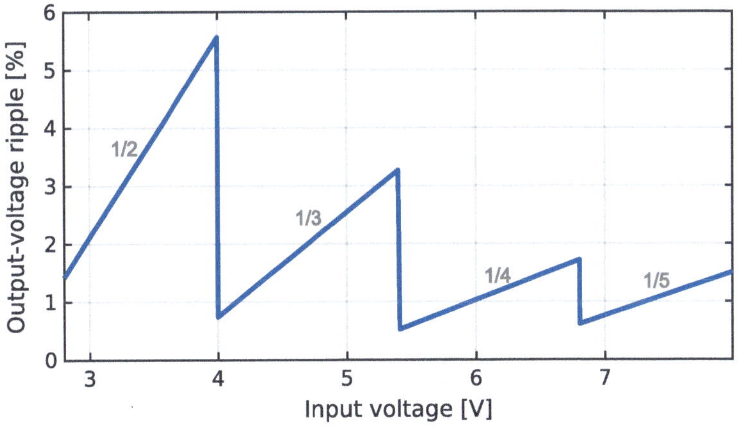

Fig. 4.4 Output-voltage ripple versus input voltage for the prior multi-ratio Dickson-star topology from [64], with iVCRs annotated

4.4 Topology

To address the two key challenges introduced by the wide input voltage range, a topology is proposed. First, Sect. 4.4.1 presents a multi-ratio topology that mitigates output impedance losses, particularly optimized for rectified AC inputs. Second, Sect. 4.4.2 further enhances this topology by a ripple reduction technique, which is suitable for high voltages. The resulting multi-ratio topology builds upon the single-ratio topology presented in Chap. 3, which has been theoretically proven optimal for high voltage. Note that the techniques described in this section are not limited to this specific topology and can be applied to other SCC topologies as well; hence, they are presented for general SCC stages.

4.4.1 Mitigating Output Impedance

Enlarging the configuration switches reduces their parasitic resistance and, hence, the converter output impedance. Yet, this adjustment leads to more switch-parasitic losses. Fortunately, the switch-parasitic losses caused by the reconfigurations remain negligible for a rectified AC mains input since they occur only once per rectified mains period (~100 Hz) for every switch. Nevertheless, there is another source of switch-parasitic losses because of the connection of these configuration switches to the flying capacitors. The voltage swings of these flying capacitors induce drain-source voltage swings (ΔV_{ds}), which periodically charge the large drain-source capacitances (C_{ds}) at the SCC switching frequency (up to 42.5 MHz in this design). Consequently, upsizing the configuration switches shifts the problem toward augmented switch-parasitic losses.

To mitigate these switch-parasitic losses due to the flying capacitor swings ($P_{sw,SCC}$), this work introduces a technique where the configuration switches are connected to nodes that do not swing along with the flying capacitors (Fig. 4.5). These DC nodes are created at each stage's input and output, as will be discussed in Sect. 4.5.1. Reconfiguring DC nodes allows for large configuration switches while maintaining negligible switch-parasitic losses, addressing the initial challenge of high parasitic resistance. However, note that the area occupied by these larger switches becomes substantial. This introduces a new trade-off between the on-resistance and area of the configuration switches.

Figure 4.6 depicts how the DC nodes are interconnected to realize multiple topologies. The interconnections of the stage inputs and outputs initially form a cascade configuration at the highest SCC input voltage. As the SCC input voltage lowers, the stages are gradually reconfigured in parallel. Stages are initially paired, and subsequently these pairs are further grouped into clusters of four stages, which are then further expanded to clusters of eight stages. This iterative process continues until all stages are configured in parallel. This gradual parallelization reuses all available capacitance, leading to enhanced performance.

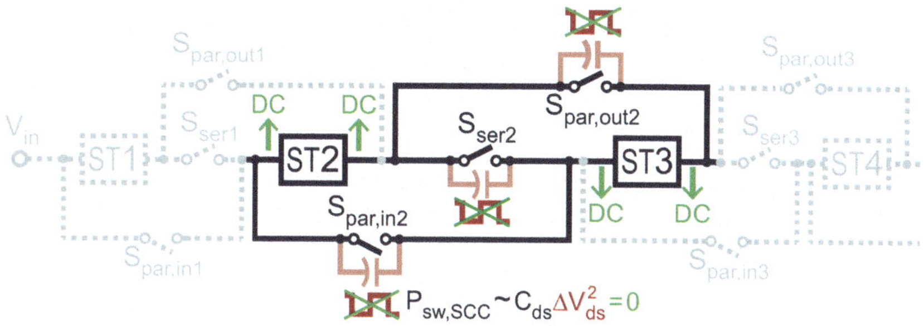

Fig. 4.5 Configuration switches (S_{par} and S_{ser}) connect generated DC nodes at the input and output of the stages (ST), which eliminates their switch-parasitic losses ($P_{sw,SCC}$)

Two clusters are combined by connecting the respective inputs and outputs of the neighboring stages. This is shown in Fig. 4.7a, where two clusters of four stages are combined by interconnecting the respective inputs and outputs of ST4 and ST5. This arrangement allows for configuration switches with a low voltage rating (V_{sw}) since adjacent stages have the lowest voltage difference throughout the AC period. Conversely, the same two clusters could be combined to attain an identical conversion ratio by connecting the inputs of ST1 and ST5, and by connecting the outputs of ST4 and ST8. While this alternative decreases the charge passing through the configuration switches, it requires higher voltage ratings for the transistors. These higher ratings require immense termination structures, as previously explained in Sect. 3.3.1. This results in a less advantageous trade-off between area and losses, with the latter determined by the FSL output impedance (2.17).

Gradual parallelization realizes various topologies with a distinct number of stages in cascade (F), each associated with an iVCR described by

$$iVCR = \frac{1}{1+F}. \tag{4.3}$$

Collectively, these configurations form the following set of iVCRs:

$$iVCRs = \left\{\frac{1}{2}, \frac{1}{3}, \ldots, \frac{1}{N_{max}}\right\}. \tag{4.4}$$

These iVCRs are selected to optimize the efficiency for the predefined rectified AC wave, as demonstrated in Fig. 4.8. The top graph shows the SCC input voltage across one period, while the bottom graph represents the corresponding theoretical ideal efficiency, accounting only for intrinsic losses. Importantly, the input voltage spends most of its time at high voltage because of its sinusoidal shape. Consequently, the iVCRs are chosen to be the most efficient during these extended high-voltage intervals, thereby enhancing overall efficiency. While efficiency dips arise at low input voltages, their impact is minimal because of their brief duration within an AC cycle.

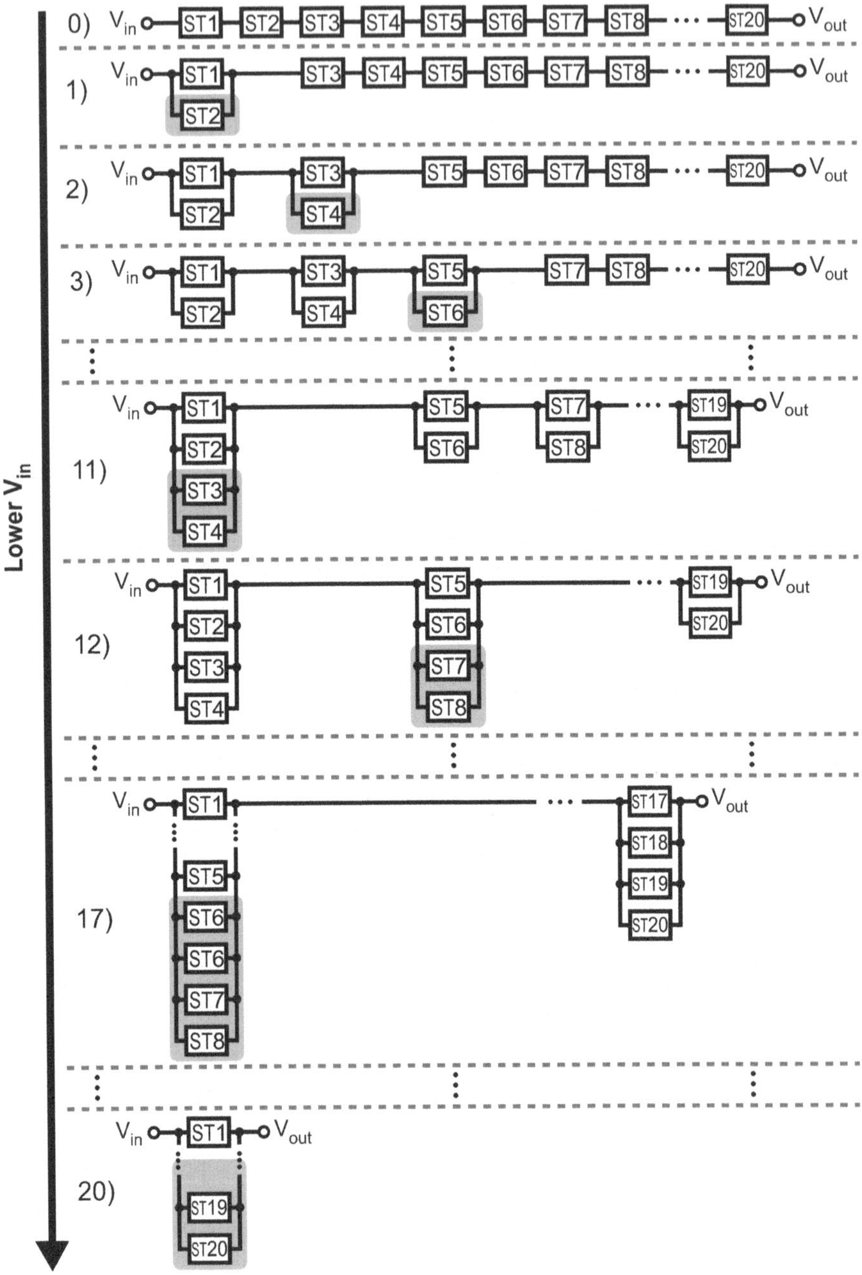

Fig. 4.6 The SCC stages (ST) are initially cascaded at the highest SCC input voltage (V_{in}). They are gradually grouped in parallel as V_{in} decreases

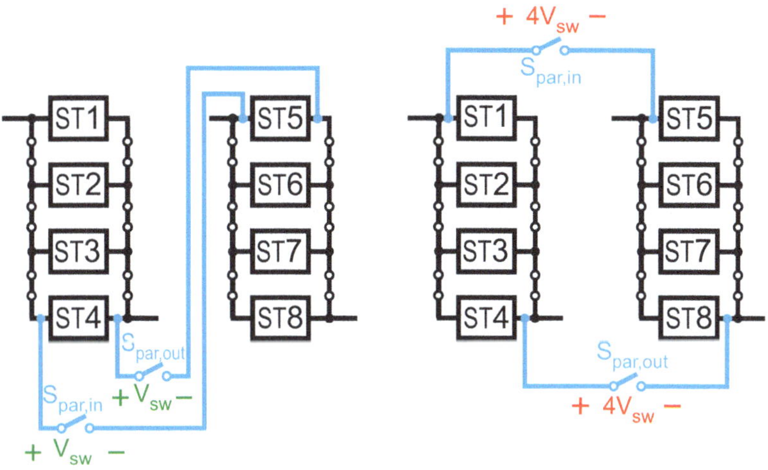

(a) (b)

Fig. 4.7 Combining two clusters of four stages by interconnecting (**a**) the adjacent stages or (**b**) the respective cluster inputs and cluster outputs

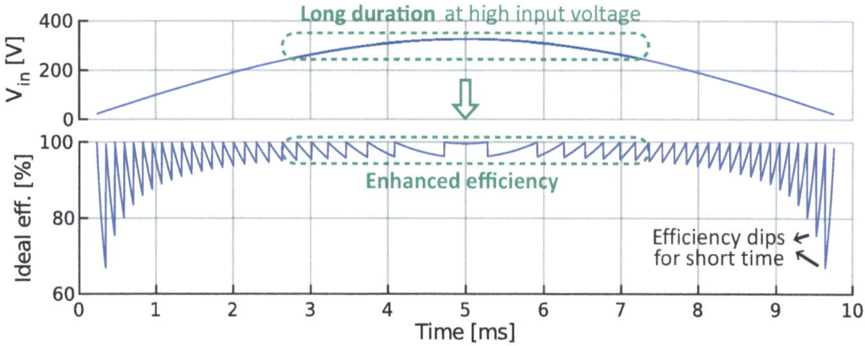

Fig. 4.8 Rectified AC input voltage (top) and corresponding SCC ideal efficiency (bottom) versus time for one period

4.4.2 *Ripple Reduction for Multi-ratio SCCs*

While the proposed multi-ratio topology successfully tackles the first challenge of increased output impedance, it does not mitigate the excessive output-voltage ripple. Before delving into the solution, it is necessary to provide more details about the underlying cause of the increased ripple. Multi-ratio topologies exhibit a high output-voltage ripple (ΔV_{out}) for specific input voltages, as shown by the blue curve in Fig. 4.9. These particular input voltages correspond to low switching frequencies, which are inversely proportional to the maximum output-voltage ripple [81]:

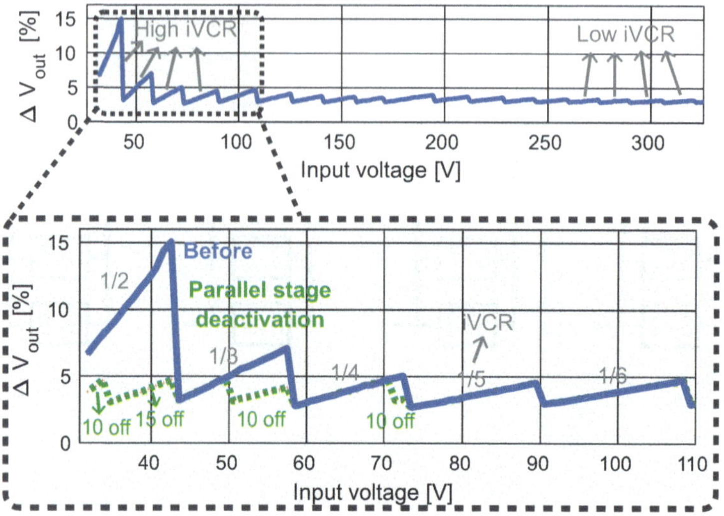

Fig. 4.9 Multi-ratio topologies exhibit high output-voltage ripple (ΔV_{out}) at certain input voltages (blue). The measured steady-state ripple is reduced in the dotted green graph during high ripple peaks by parallel stage deactivation. The iVCRs are annotated in gray

$$\Delta V_{out} = \frac{I_{out}}{2 \cdot f_{sw} \cdot M \cdot (C_{out} + \kappa_\tau \cdot C_{fly})}, \tag{4.5}$$

where I_{out} represents the load current, M denotes the number of interleaved fragments, and κ_τ indicates the amount of flying capacitance functioning as additional decoupling. Low switching frequencies and thus elevated ripple occur when two conditions are met. The first condition is a topology with a high iVCR,[1] as they require a lower charge transfer through their flying capacitors than those with a low iVCR. Hence, the switching frequency decreases for a specified output power (cf. Eq. (2.14)). The second condition is a high input voltage for a particular iVCR. To accommodate this increased input voltage, where the effective voltage conversion ratio (VCR) deviates from the iVCR, the residual voltage conversion drops across the output impedance. This necessitates a larger output impedance, established by reducing the switching frequency.

To decrease the ripple, conventional techniques occupy too much area for high-voltage SCCs. Time interleaving [67, 82] divides components into smaller units that operate out of phase, leading to significant isolation spacings between the numerous units at high voltage. Likewise, capacitance modulation [5, 27, 80] splits the flying capacitors into smaller units that can be deactivated, thereby facing the same spacing problem. Other ripple reduction techniques modulate the equivalent resistance of

[1] A high iVCR refers to a numerically large value. For instance, an iVCR of 1/2 is considered higher than 1/20 in this context.

the power switches, through adjustments in their width [24], duty cycle [27], or gate voltage [24]. Switch width modulation needs segmentation into smaller power switches, reintroducing spacing-related challenges. While tuning the gate voltage or duty cycle does not inherently demand such segmentation, their effectiveness is restricted to the FSL region. To maintain operation within the FSL region across the entire power range, additional modulation of the equivalent resistance is necessary. Achieving this by solely adjusting the gate voltage or the duty cycle is unfeasible due to their limited turning range. Consequently, it becomes imperative to complement these two methods with earlier approaches involving component segmentation, requiring extensive isolation area once again.

As a solution, this work presents *parallel stage deactivation*, which deactivates certain parallelized stages during periods of high ripple. This is illustrated in Fig. 4.9, where the original ripple versus input voltage is depicted in blue. Within this figure, the green dotted line reduces the ripple by deactivating stages during intervals of elevated ripple, with the number of disabled stages annotated. Beyond an iVCR of 1/4, stage deactivation becomes unnecessary, as the ripple was already consistently low before within that region. Note that at the lowest input voltage, stages are immediately deactivated. This is because, at this specific iVCR value (i.e., 1/2), there is a very narrow voltage range within which the ripple remains acceptable without the need for stage deactivation.

The corresponding topologies for iVCR values of 1/2 and 1/3 are shown in Fig. 4.10, highlighting the parallel stages that are disabled. These parallel stages are readily available for the given high iVCR, thanks to the gradual parallelized topology that puts stages in parallel when going toward higher iVCRs. These parallel stages are also available for other iVCRs that incorporate stage deactivation, as deactivation is solely employed at high iVCRs where the ripple is high, as specified by condition 1. Consequently, this eliminates the need for component segmentation in contrast to state-of-the-art approaches, avoiding excessive area overhead at high voltage.

By disabling stages, the active flying capacitance decreases, requiring a proportional rise in switching frequency to maintain the same output power. By substituting f_{sw} with $S_{on,rel}$, while reducing the active flying capacitance by a factor of $S_{on,rel}$ in (4.5), the resulting output-voltage ripple is described by

$$\Delta V_{out} \propto \frac{S_{on,rel}}{C_{out} + \kappa_T \cdot (S_{on,rel} \cdot C_{fly})}, \tag{4.6}$$

where $S_{on,rel}$ represents the ratio of active to total stages. The denominator in (4.6) accounts for the flying capacitors of the deactivated stages, which no longer provide additional decoupling. However, this effect can be neglected if the corresponding decoupling capacitance is small in comparison to C_{out}. This simplifies (4.6) to

$$\Delta V_{out} \underset{\sim}{\propto} S_{on,rel}. \tag{4.7}$$

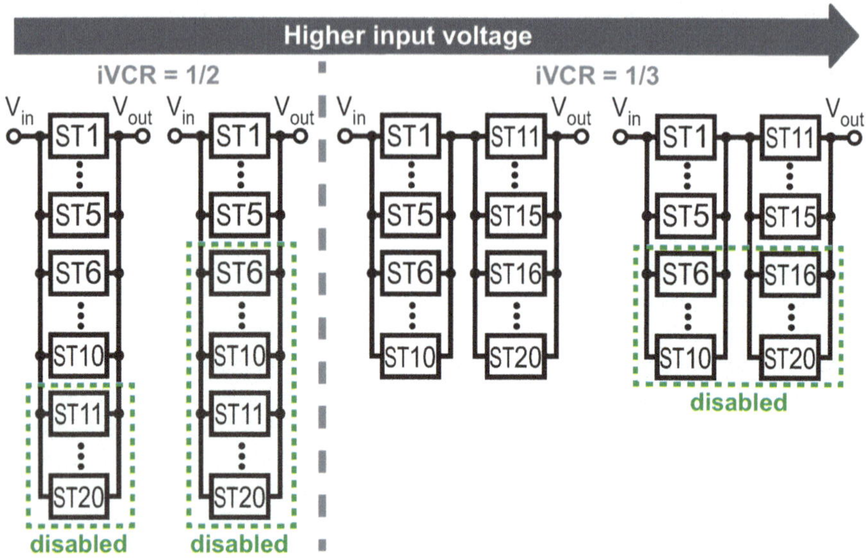

Fig. 4.10 Deactivation parallelized stages for iVCR values of 1/2 and 1/3

This approximation is valid for the current design. To illustrate, when $S_{on,rel} = 0.5$, the approximated formula (4.6) yields a ripple reduction of 2×, closely matching 1.93× by the precise formula (4.7). In designs where this effect is nonnegligible, it can be mitigated by connecting the deactivated flying capacitors to the output, thereby serving as output decoupling, rather than disconnecting them.

In addition to reducing steady-state ripple, this technique also decreases the output-voltage overshoot during topology transitions by the same factor. When reconfiguring two stages from cascade to parallel, their flying capacitors are put in parallel. As these flying capacitors operated at different voltages before the transition, this results in charge redistribution. This causes charge to flow to the output, thereby inducing an overshoot. Deactivating parallel stages during topology transitions reduces the overshoot, as there are fewer flying capacitors redistributing charges.

The stage deactivation scheme of Fig. 4.9 already realizes this overshoot reduction. When the input voltage decreases (moving from right to left in Fig. 4.10), stages are deactivated at critical topology transitions to reduce overshoot. Afterward, when the input voltage further decreases following the topology transition, stages are reactivated, while the topology is not reconfigured. While this reactivation introduces an additional overshoot, the overall overshoot magnitude reduces since the charge transfer is smeared out over time. Conversely, when transitioning from parallel to cascade (moving from left to right in Fig. 4.10), parallel stages are activated at topology transitions. Both the reconfiguration and the activation at increasing input voltages extract charge from the output. This output discharge is counterbalanced by an additional charge transfer phase of the SCC (initiated by the output control), thereby preventing voltage droop.

Parallel stage deactivation does not impact the maximum achievable output power since stages are exclusively deactivated for input voltages where $VCR << iVCR$ (see condition 2). In this scenario, the increased voltage difference between the flying capacitors ($\propto (iVCR - VCR)$) results in a greater charge transferred each cycle. This augmentation counterbalances the reduction in output power (P_{out}) caused by the decreased flying capacitance ($C_{fly,tot}$). This can be mathematically expressed by reformulating (3.19):

$$P_{out} \propto f_{sw} \cdot C_{fly,tot} \cdot \big(V_{in} \cdot (iVCR - VCR)\big). \tag{4.8}$$

Moreover, since the converter can achieve the maximum output power at all input voltages despite the C_{fly} reduction, it facilitates rapid transitions from light to full load. During these load steps, there is no need to activate additional C_{fly} units. This contrasts with prior studies that use capacitance modulation [5, 27, 80], which sense the output voltage or switching frequency instead of the input voltage for their control. Their gradual activation of additional C_{fly} units via a slow secondary control loop results in a voltage droop during load steps.

Next, the impact of parallel stage deactivation on efficiency is examined. The power losses in SCCs comprise the slow-switching limit losses (P_{SSL}), the fast-switching limit losses (P_{FSL}), the switch-parasitic losses ($P_{par,sw}$), and the capacitor-parasitic losses ($P_{par,C}$), as earlier discussed in Sect. 2.2.[2] With stage deactivation, both the flying capacitances and the power switch widths are divided by $S_{on,rel}$, while the switching frequency is multiplied by the same factor. Using the formulas from Sect. 2.2, the resulting power losses (P_{loss}) are calculated as follows:

$$P_{loss} = P'_{SSL} + \frac{P'_{FSL}}{S_{on,rel}} + P'_{par,sw} + P'_{par,C}, \tag{4.9}$$

where an apostrophe denotes the original loss term without stage deactivation. All loss terms remain constant, with the exception of P_{FSL}. The influence of the rising P_{FSL} depends on its magnitude relative to the other terms. Since stages are deactivated only when $VCR << iVCR$ (see condition 2), where the SSL output impedance is raised to accommodate the lower VCR, P_{FSL} is generally minor compared with P_{SSL}. Hence, the overall impact is limited. Moreover, since stages are only disabled during the 4% of the AC period (i.e., when the ripple was high before), the effect on the overall efficiency is minimal.

[2] The converter loss calculation approximates the Euclidean norm of the SSL and FSL output impedance losses of (2.25) by a linear addition. This yields a straightforward and intuitive analytic expression, without significantly deviating from the actual result [8, 30].

4.5 Implementation

This section delves into the converter implementation, beginning with the multi-ratio SCC. Next, it introduces drivers that reduce the area and the losses. Lastly, the input control is discussed, which encompasses efficient high-voltage input sensing and a specialized ADC.

4.5.1 *Multi-ratio SCC*

Section 3.5.1 provides an in-depth explanation of the DC node creation [8]. Figure 4.11a–c demonstrates the introduction of stages with DC inputs and outputs during this generation process, employing a Dickson-star topology for illustration.[3] The partitioning into stages is realized by splitting the interconnecting power switch between two flying capacitors (1.9 pF, rated for 325 V) into two power switches, as depicted in Fig. 4.11b. This forms Dickson-star stages operating in a cascade configuration. These stages are then interleaved four times (0°, 90°, 180°, and 270°), thereby generating DC nodes at both their inputs and outputs (Fig. 4.11c).

These DC stage inputs and outputs are reconfigured by a multi-ratio configuration block, as conceptually shown in Fig. 4.11d. Figure 4.12 depicts the implementation of the multi-ratio SCC topology, including this multi-ratio configuration block and the stages. The configuration switches connect only to DC nodes, which eliminates the switch-parasitic losses due to the flying capacitor swings, as was proposed in Sect. 4.4.1. Furthermore, the configuration switches only interconnect neighboring stages, facilitating a low-voltage implementation (rated for 32 V, 29 Ω each).

In the stages, the nMOS and pMOS top power switches are rated for 32 V and have on-resistances of 1.1 kΩ and 2.9 kΩ, respectively. The nMOS and pMOS bottom power switches are rated for 18 V and have a 54 Ω and 144 Ω on-resistance, respectively. The bottom power switches have a lower resistance than the top power switches to compensate for their larger charge flow.

The multi-ratio topology enables the implementation of the sidewall coupling reduction technique introduced in the previous chapter. In particular, the clock phase of all power switches remains constant across all topologies (cf. Fig. 4.12), resulting in flying capacitors with a fixed voltage swing. By placing flying capacitors with matching swings together in the layout, as detailed in Chap. 3, their mutual sidewall coupling is eliminated. To allow this constant phase across all topologies, the antiphase fragments ensure charge balance in both phases. The absence of antiphase fragments would lead to an imbalance of charge flowing in and out of the intermediate DC nodes, as illustrated in Fig. 4.13a. Stage 1 outputs charge during phase Φ_1, while stage 3 does not accept any charge during that phase. Figure 4.13b

[3] Other SCC topologies can also be transformed into stages with DC nodes, as described in [8].

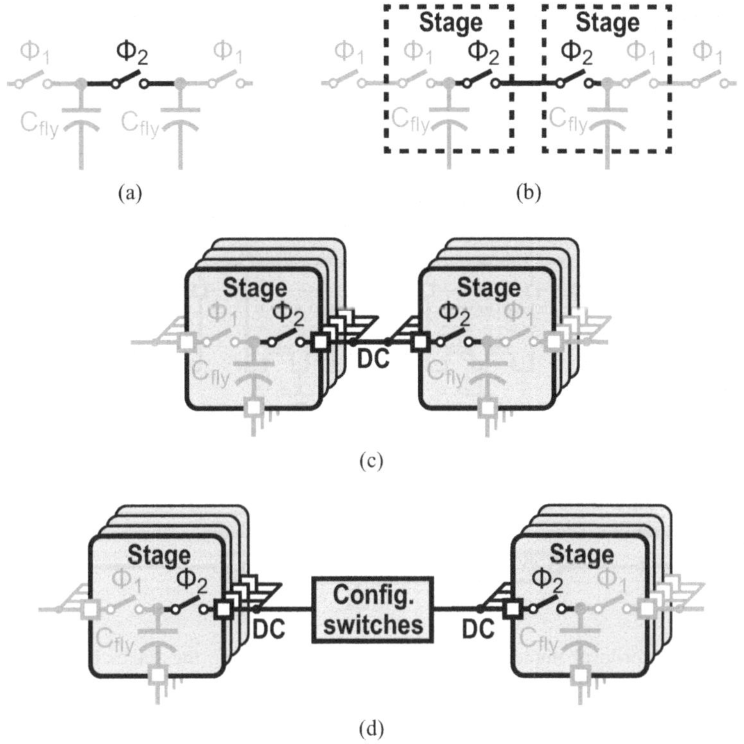

Fig. 4.11 Conceptual representation of going from (**a**) the top part of a Dickson-star topology to (**b**) split interconnecting power switches, to (**c**) stage interleaving, to (**d**) reconfiguration of stage inputs and outputs

illustrates how the out-of-phase fragments resolve this issue. The 180°-shifted stage 3 accepts the output of the 0°-shifted stage 1 during phase Φ_1.

A lower-bound hysteretic control regulates the output voltage clocked at 170 MHz ($f_{clk,out}$), similar to the one in Sect. 3.5.4.

4.5.2 Drivers

Both the configuration and power switches have their own specialized driver types. This subsection proposes charge delivery and level-shifter circuits that minimize the area and the losses.

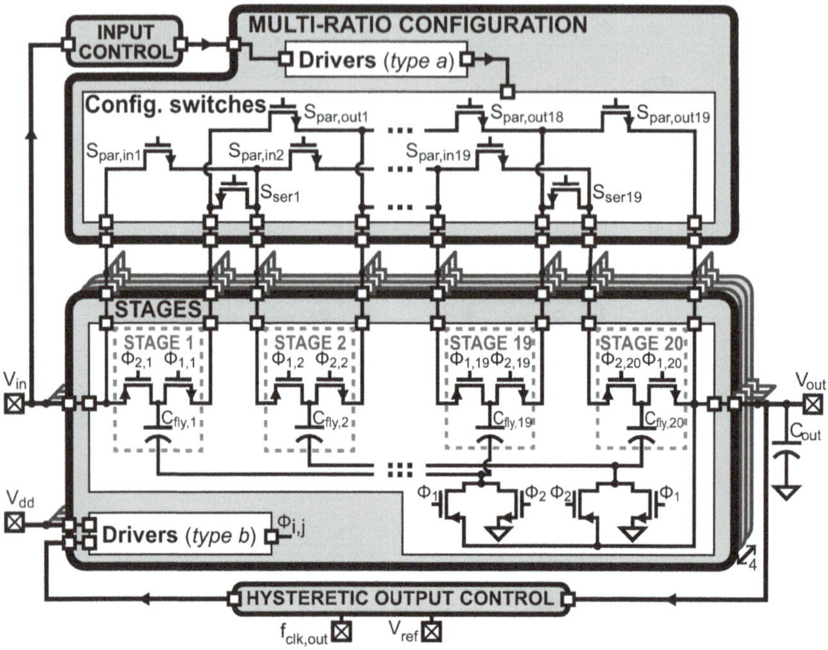

Fig. 4.12 Transistor-level implementation of the SCC, including the stages (bottom) and the multi-ratio configuration (top)

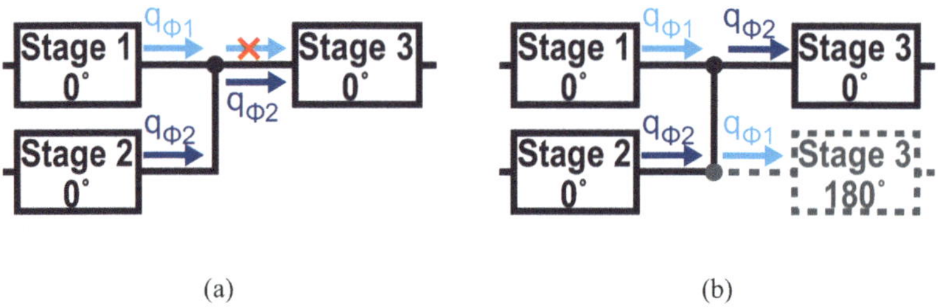

Fig. 4.13 Example of charge flow (**a**) without the antiphase fragments and (**b**) with the 180°-shifted stage 3

4.5.2.1 Charge Delivery

The activation of the configuration switches necessitates charge delivery to their parasitic gate-source capacitor (C_{gs}) at high voltage (driver *type a* in Fig. 4.12). However, the sizable C_{gs} would result in considerable driver area overhead for each of the 57 configuration switches when using state-of-the-art techniques. In [15], the charge is delivered by the supply voltage (V_{dd}) through a level-shifter, which requires a large high-voltage level-shifter capacitor (C_{LS}) to mitigate the voltage

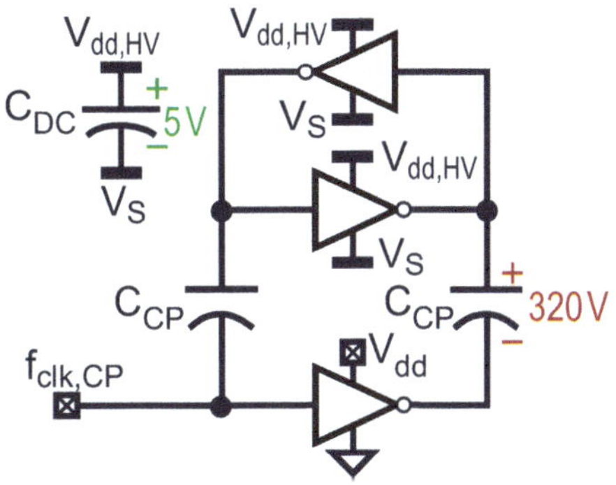

Fig. 4.14 Charge pump implementation of the configuration switches

drop ($\propto \frac{C_{LS}}{C_{gs}+C_{LS}}$). Alternatively, another approach [35] employs a charge pump, which also leads to a substantial area penalty. Specifically, as the source voltage of the driven switch does not return to 0 V and varies $> V_{dd}$ during topology changes, a complex self-boost charge pump is needed. This charge pump demands four sizable high-voltage components with extensive high-voltage termination structures and suffers from three diode voltage drops.

To minimize the area overhead and the diode voltage drops, this work introduces the charge pump shown in Fig. 4.14 to generate the supply at high voltage ($V_{dd,HV}$) from the corresponding switch source voltage (V_S). The circuit is based on the level-shifter in [15] but functions as a dedicated charge pump in this work. By clocking ($f_{clk,CP}$) the charge pump 7500× faster (1.5 MHz) than the associated level-shifter, the high-voltage capacitance (65 fF) and thus the area are substantially reduced. While the charge pump gradually charges the decoupling capacitor (C_{DC}) in the high-voltage domain, C_{DC} must instantaneously provide the gate charge during activation. Therefore, the proposed charge pump necessitates its C_{DC} (5 pF) larger than C_{gs} to minimize the voltage drop ($\propto \frac{C_{DC}}{C_{gs}+C_{DC}}$), similar to the charge delivery by the level-shifter in [15]. In contrast, the associated voltage rating is V_{dd} (5 V) and thus 65× smaller than in [15], which decreases its area proportionally based on the scaling trend (3.6) that was derived in Sect. 3.4.1.2. Additionally, the proposed charge pump eliminates diode voltage drops, and it only requires two high-voltage capacitors (C_{CP}), which do not need large termination structures like the high-voltage switches in [35] (cf. Sect. 3.3.1).

Although the preceding charge delivery method is tailored for a large switch switching at a low frequency, the power switches necessitate an alternative driver design. The power switches are small and operate at a varying high frequency (driver *type b* in Fig. 4.12). Each switch is small since it handles a small portion of the total charge; however, the combined area of the 160 power switches and their drivers becomes significant. To optimize this driver area, the charge is supplied by the level-shifter from [15]. Incorporating a charge pump would not decrease the area, since its components require additional spacings for high-voltage isolation. These spacings dominate the area because of the small component sizes in each driver (cf.

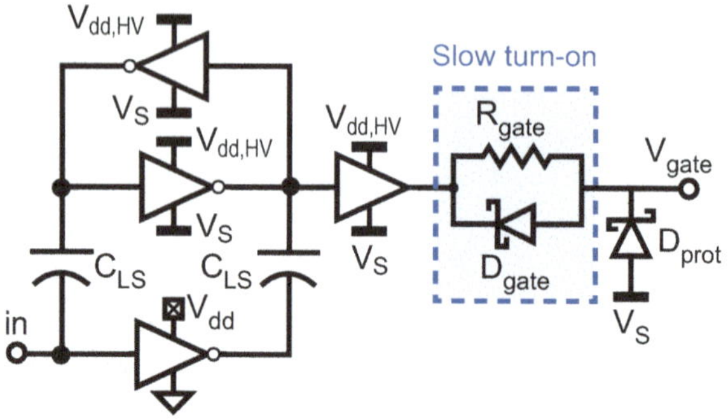

Fig. 4.15 Level-shifter implementation of the configuration switches

Sect. 3.3.1). In addition, the implemented approach also minimizes the switching losses of the drivers, which are significant due to the high operating frequency. The level-shifter switches only when charge is needed thanks to the hysteretic control, which scales its switching losses with output power.

4.5.2.2 Level-Shifters

Similar to the charge pump, the conventional level-shifters [36, 60] occupy too much area. The problem arises from the fast source voltage variations caused by topology changes, which requires complex auxiliary circuits as in [36] for each of the 57 configuration switches and the 160 power switches. In contrast, this work addresses this issue by intentionally slowing down the topology transitions. Hence, the level-shifter in Fig. 4.15 slowly turns on its configuration switch by incorporating a series resistor (R_{gate}) in the gate driving path. This method allows for more compact capacitive level-shifters for all drivers, eliminating the need for complicated circuits to accommodate rapid dV/dt changes.

The series resistor only delays the activation of the configuration switch. Conversely, during deactivation, a parallel Schottky diode (D_{gate}) establishes a low-impedance path between the gate and the source of the power switch.[4] This is crucial when the drain voltage rises due to another power switch, causing coupling to the gate through the parasitic gate-drain capacitor. The low-impedance path provided by the Schottky diode prevents an unintentional rise in the gate voltage, thereby ensuring the transistor remains off as intended.

Finally, the diode D_{prot} protects the gate from under-voltage conditions. Meanwhile, the over-voltage protection is already provided by D_{gate}.

[4] Schottky diodes are typically available in high-voltage CMOS technologies.

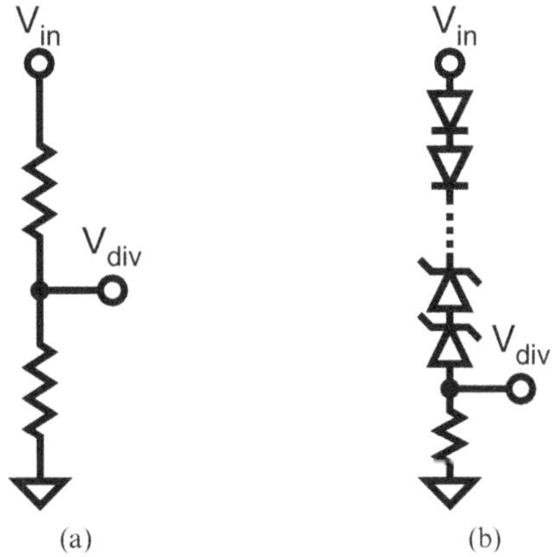

Fig. 4.16 Conventional input sensing: (**a**) resistive division and (**b**) (Zener) diode stack

4.5.3 *Input Control*

The input control determines the SCC topology based on its input voltage. For this purpose, a voltage divider down-converts the input voltage range (0–325 V) to fit within the ADC supply rails (0–5 V). Following this voltage division, an ADC converts the divided input voltage into a digital state. This state maps into the iVCR and the number of deactivated stages.

4.5.3.1 Input Sensing

State-of-the-art approaches (Fig. 4.16) cannot achieve both an accurate and low-power voltage division. Conventional resistive voltage division [15] is unsuitable because they lead to tremendous power consumption at the high operating voltage (e.g., 53 mW for 1 MΩ at 230 V_{RMS}). On the other hand, relying on the threshold voltages of a (Zener) diode stack [60] would introduce a strong dependence on process and temperature. This dependence necessitates substantial overdesign to prevent output voltage drops, which would significantly degrade efficiency and increase the buffer capacitor area for AC-DC conversion.

Capacitive dividers could eliminate the high power consumption while achieving a high accuracy. However, conventional capacitive dividers cannot deal with DC inputs (or with the slow AC mains input), as leakage sets the divided output voltage (V_{div}). Therefore, this work proposes a resettable capacitive divider, as illustrated in Fig. 4.17 with capacitors C_{d1} (94.5 fF) and C_{d2} (10.9 pF). The divided output is biased with a bias voltage (V_{bias}) during a short reset phase (Φ_{rst}), as shown in Fig. 4.17b. For the remaining time, a capacitive divider is connected between the input and the output in phase $\overline{\Phi_{rst}}$ (Fig. 4.17c). This connection enables the output voltage to track the input voltage according to the capacitor divider ratio. The resulting output voltage can be expressed as

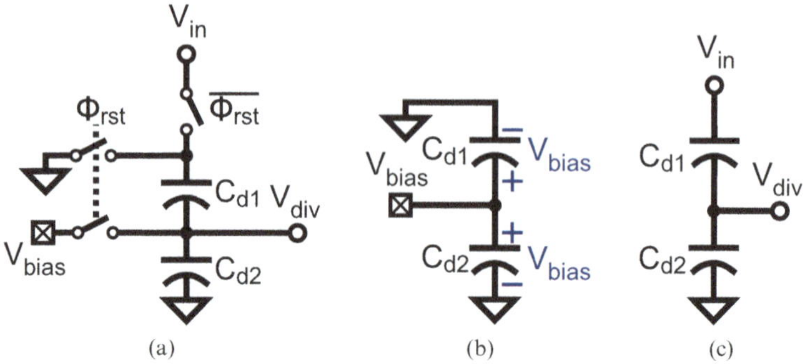

Fig. 4.17 Resettable capacitive divider: (**a**) schematic, (**b**) equivalent circuit in the resetting phase, and (**c**) equivalent circuit in the active mode

$$V_{div} = V_{bias} + \frac{C_{d1}}{C_{d1} + C_{d2}} \cdot V_{in} - V_{leak}(t), \tag{4.10}$$

where V_{leak} represents the voltage reduction over time (t) due to leakage. To counteract this leakage, the bias voltage is periodically reset by phase Φ_{rst} at a rate of 5 kHz. These resets occur in between two ADC clocks, ensuring that the ADC always receives a valid input signal at the moment of conversion.

Although the resettable capacitive divider may resemble a switched-capacitor resistor, their operation and losses are different. Switched-capacitor resistors lack a direct electrical path between the input and the output, requiring constant switching for the divided output to follow the input. This high switching frequency imposes substantial losses, especially when the capacitor is charged to a high input voltage, resulting in losses that scale quadratically with this voltage. In contrast, the proposed resettable capacitive divider continuously tracks the input voltage throughout phase $\overline{\Phi_{rst}}$, not requiring constant switching to maintain this connection. In this case, the switching action compensates for leakage, requiring a considerably slower frequency and, thus, leading to a more efficient solution.

Three small implementation techniques optimize the resettable capacitive divider. First, the parasitic substrate capacitors of C_{d1} and C_{d2} are designed to be at the top plate of C_{d1} and the bottom plate of C_{d2}, instead of at node V_{div}. This placement prevents the parasitic capacitors from being in parallel with C_{d2}, which would otherwise alter the division ratio and deteriorate matching. Second, the ADC capacitances at node V_{div} are optimized, as they are taking part in the capacitive division. The ADC sample-and-hold capacitors implement the same capacitor type as used for C_{d1} and C_{d2} (sandwich, rated for 325 V) to improve matching. Additionally, during the reset phase, the ADC input capacitances are also charged to V_{bias} as they are in parallel with C_{d2}, further improving the accuracy. Third, the top reset switch in the divider is opened after the bottom reset switch at the end of the reset phase. This sequencing mitigates clock feedthrough by the top reset switch to enhance the accuracy.

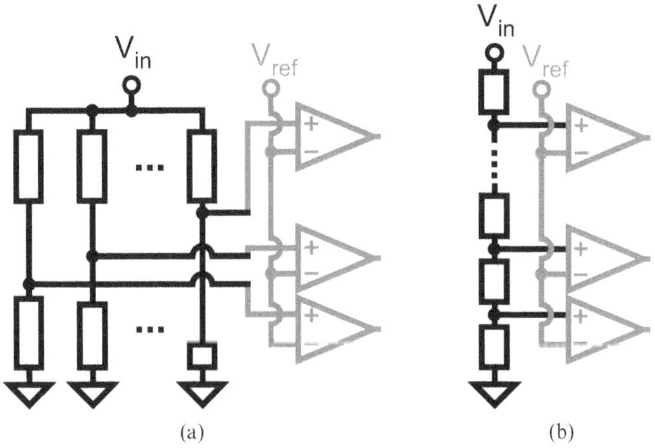

Fig. 4.18 Conventional flash ADC with (**a**) multiple voltage dividers having distinct ratios and (**b**) a single voltage divider with multiple output taps

4.5.3.2 Specialized ADC

Following the voltage division, conventional works use flash ADCs [15, 17, 78], as shown in Fig. 4.18. However, they prove impractical for input control at high voltage. In these prior works, every iVCR requires a distinct voltage division of the input signal. This ensures that the divided voltage falls within the common-mode range of the associated comparator at the iVCR threshold. The methods for realizing these diverse division ratios face significant challenges when dealing with high voltages. The first approach [14] employs a separate divider for every iVCR with an optimized voltage ratio, as depicted in Fig. 4.18a. However, this consumes excessive area due to the numerous iVCRs required in this design, each requiring large high-voltage capacitors. The second approach [17, 78] relies on multiple output taps of a single voltage divider, as illustrated in Fig. 4.18b. All these output taps suffer from parasitic capacitances in their interconnects, which are significant compared to the relatively low capacitance of the high-voltage capacitors. This results in an alteration of the division ratio and poor matching.[5]

To minimize the area usage, this work implements a single voltage divider with just one output tap connecting to the ADC. Using this single divider for prior resistive or diode dividers would divide their low input voltages to levels near 0 V,

[5] Note that a previous work [63] solves the need for multiple high-voltage dividers by implementing output control instead of the traditional input control for ratio control. However, this approach involves periodic experimentation with multiple iVCRs to determine the optimal one, resulting in an increased ripple during these testing phases. Furthermore, this method selects the lowest iVCR setting at which the output voltage remains above the reference voltage at the maximum frequency. This would lead to suboptimal losses and a slow transient response in this design when operating at low power.

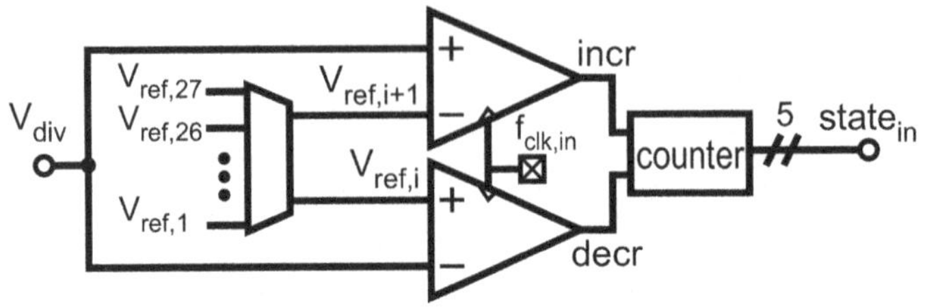

Fig. 4.19 ADC specialized for the continuous input voltage shape

having them outside the common-mode voltage range. In contrast, the proposed resettable capacitive divider (Fig. 4.17) introduces an additional design variable V_{bias}. This parameter offsets the divided input voltage (4.10), thereby enabling to shift the low divided input voltages within the ADC common-mode voltage range. Furthermore, to accommodate the high input voltages within the common-mode range as well, the divider ratio is selected accordingly.

Following the single voltage divider, the ADC is optimized for the specific AC input voltage shape (Fig. 4.19). This input voltage demonstrates smooth and continuous changes up to 4 MV/s, as detailed later in Sect. 5.3. This necessitates only two comparators (StrongARM [56]) clocked at 4 MHz ($f_{clk,in}$) comparing the ADC input with the voltage references ($V_{ref,i}$) just above and below the preceding value. If the current value is above or below the respective value, the counter is incremented (incr) or decremented (decr) accordingly. This, in turn, updates the input state ($state_{in}$) and adjusts the two selected references to align with the most recent value. This level-crossing ADC realizes a B-bit conversion using just two 1-bit comparisons. In contrast, traditional approaches need $2^B - 1$ comparisons for a flash ADC [15], B comparisons for a SAR ADC [63], and worst-case $2^B - 1$ comparisons for a ramp ADC [4]. Hence, the level-crossing ADC leads to a more favorable trade-off between power consumption, speed, and area.

4.6 Chip Measurements

Using the introduced topology and implementation techniques above, a prototype is fabricated. This prototype converts a wide 31–325 V DC input range into a stable 12 V DC output voltage while delivering a peak output power of 100 mW. The design is fully integrated into a 180 nm high-voltage CMOS SOI technology, as shown in Fig. 4.20. The main building blocks encompass the SCC topology, containing its interleaved stages and configuration switches, and the input and output control. The marked area is not used for DC-DC conversion, as it serves to extend the DC-DC converter into an AC-DC converter, as described in the subsequent chapter. The converter has an active area of 6.1 mm^2. This includes the output capacitor (0.5 nF), which consumes 11.8% of the active area. Note that

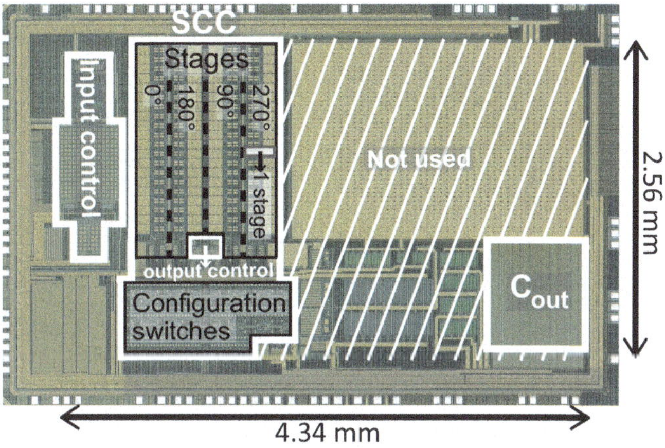

Fig. 4.20 Die micrograph of the fully integrated wide-input-range DC-DC converter

although the converter is optimized to serve in an AC-DC converter, it can also operate as a standalone wide-input-range DC-DC converter.

This section first overviews the measurement setup, after which the measurement results are presented and compared with the state of the art.

4.6.1 Measurement Setup

Figure 4.21a depicts the measurement setup, which closely resembles the one described in Sect. 3.6.1. Unlike the previous measurement setup, which relied on two multimeters and a shunt resistor to measure the input power, this one employs a dedicated power analyzer. Additionally, this setup introduces three extra clocks. While it is possible to derive these clocks from the output clock, they are externally provided to allow for greater measurement flexibility.

4.6.2 Efficiency

The efficiency measurements include all losses, except for the generation of the external clock and reference bias. Figure 4.22 shows the measured efficiency versus DC input voltage, with the corresponding iVCRs annotated. The measurements were conducted at the maximum output power of 100 mW while regulating the DC output voltage at 12 V. It realizes a 31–325 V DC input voltage range while achieving an average efficiency of 57.9% at a power density of 16.4 mW/mm^2. At high input voltages, the efficiency remains flat around its peak values thanks to the

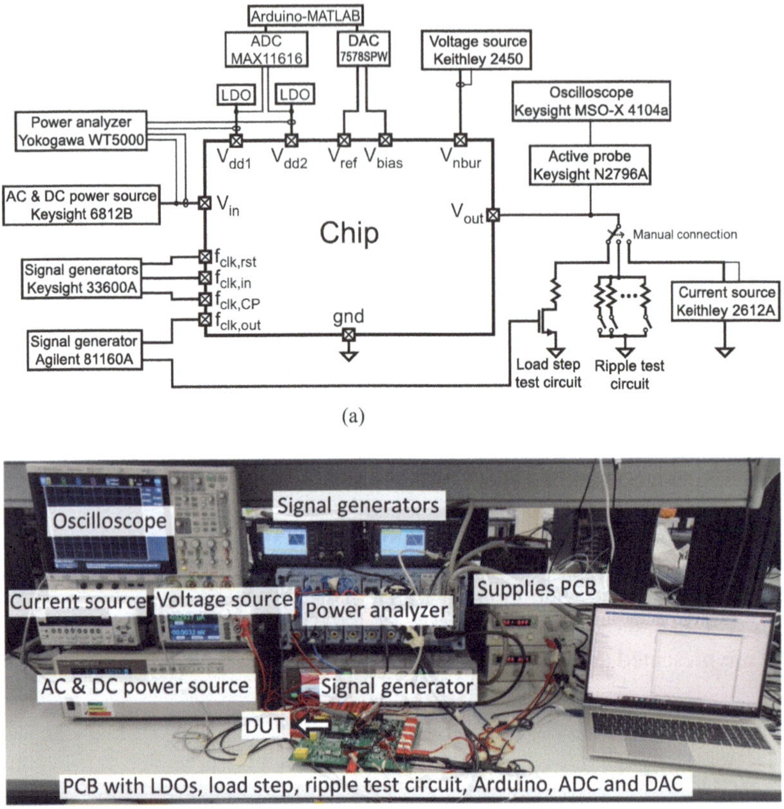

Fig. 4.21 Measurement setup. (**a**) Schematic overview. (**b**) Picture of the test equipment in the lab

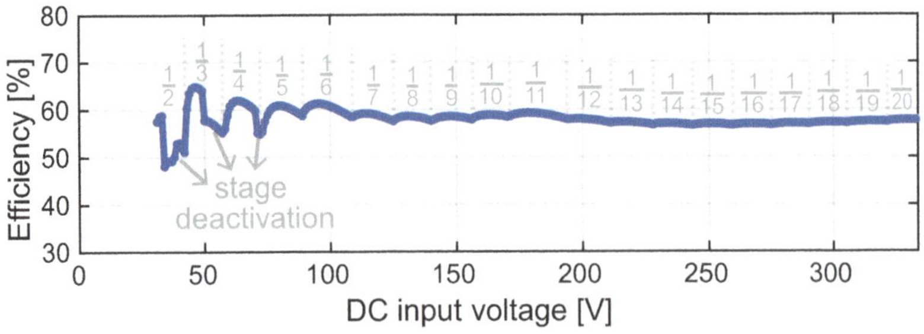

Fig. 4.22 Measured full-system efficiency versus DC input voltage for generating a DC output of 12 V at 100 mW. The iVCRs are annotated

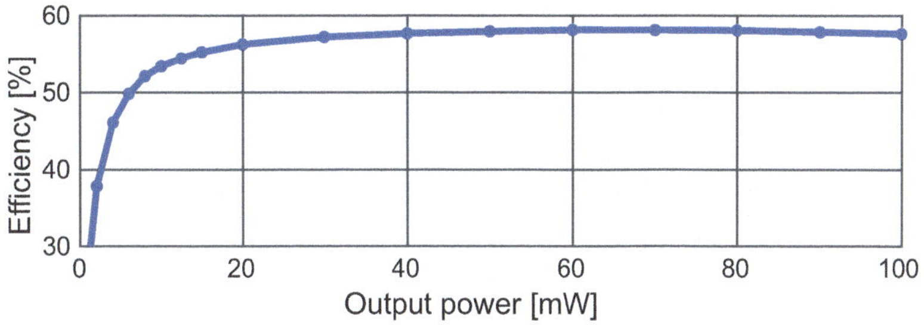

Fig. 4.23 Measured full-system efficiency versus output power for a 325-to-12 V DC-DC conversion

iVCR choice. This enhances the overall AC efficiency, as the AC input spends most of the time at these high input voltages. Conversely, the impact of the efficiency dips at low input voltages, due to stage deactivation and the iVCR selection, is limited. This is because these dips are of brief duration within an AC cycle. In particular, the measured AC-DC efficiency only drops marginally from 55.4% to 55.1% at maximum power (i.e., worst case) when enabling parallel stage deactivation.

Figure 4.23 depicts the efficiency versus output power when the DC input voltage is at its maximum value of 325 V. The efficiency curve is flat (> 53 %) within the power range of 10 mW to 100 mW, as the switching frequency scales proportionally to the output power in the hysteretic control. As the output power decreases below this range, however, the efficiency drops because the losses that do not scale with the SCC switching frequency become dominant. These fixed losses originate from various sources, including the input control, the charge pump, and the comparator in the output control, all of which operate at a constant clock frequency. Additionally, the resistive divider in the output control also incurs fixed losses since it consistently draws a constant current from the 12 V output.

4.6.3 Ripple

Figure 4.24 shows the measured steady-state output-voltage ripple versus input voltage for 0 mW output power (i.e., worst case). The blue curve represents the measured ripple when stage deactivation is enabled, while the red curve corresponds to measurements without this feature. Stage deactivation accomplishes a threefold reduction in ripple amplitude.

Furthermore, the peak-to-peak output voltage in transient, which includes overshoots at topology transitions, is 10.8%. Transient measurements are not shown here, as they will be presented in Sect. 5.5.2. The maximum value occurs when transitioning from an iVCR of 1/3 to 1/2.

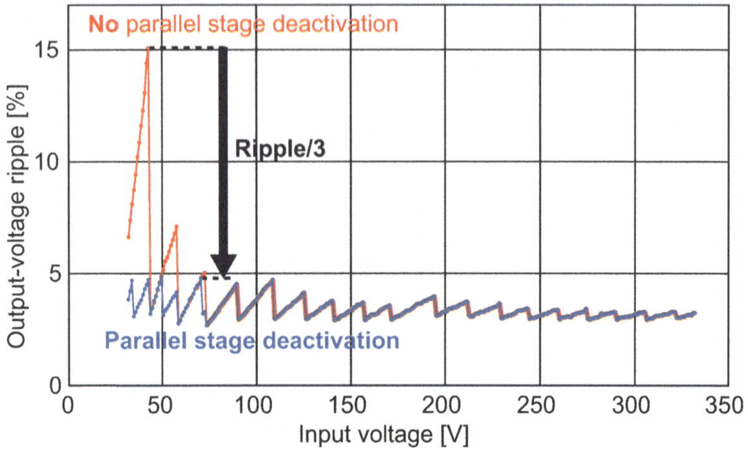

Fig. 4.24 Measured steady-state output-voltage ripple versus input voltage at 0 mW output power with parallel stage deactivation (blue) and without (red)

4.6.4 Comparison

Table 4.1 compares this work with state-of-the-art DC-DC converters with high input voltages, as well as with a linear regulator. This work eliminates the need for external components contrasting with [25, 44, 58, 60]. In addition, it achieves an efficiency substantially superior to the linear regulator [44], and a similar efficiency to [60] without the need for a bulky inductor. In comparison to previous fully integrated designs [15, 34, 85], this work achieves both a high voltage and a wide input voltage range. This work extends the maximum input voltage of fully integrated multi-ratio converters from 42 V to 325 V, while extending the iVCR range by 1.9×. In addition, it achieves an 11× higher power density at comparable efficiency.

4.7 Conclusion

This chapter targeted an SCC handling a wide DC input voltage range at high voltage toward AC-DC conversion. The proposed multi-ratio topology addressed the two key challenges: the increased output impedance and the excessive ripple. The output impedance was decreased through multiple topology techniques: reconfiguring DC nodes, gradual parallelization, low-voltage configuration switches, and optimized iVCRs. Furthermore, the ripple was reduced by deactivating parallelized stages available within the proposed topology, thereby avoiding excessive isolation spacings at high voltage. To implement the converter, efficient high-voltage input sensing and compact drivers were introduced.

Table 4.1 Comparison with state-of-the-art and commercial DC-DC converters with high input voltages

	Fully integrated				Not fully integrated			
	This work	JSSC20	ISSCC16	JSSC23	JSSC22	LR645	CICC22	ISSCC20
		[15]	[34]	[85]	[60]	[44]	[58]	[25]
Technology [nm]	180, SOI	350	350	180, SOI	180, SOI	/	180, SOI	GaN
Tech. V_{max} [V]	400	42	/	400	400	/	400	650
Topology	SCC	SCC	SCC	SCC	Resonant + buck	Linear regulator	Flyback	Buck
Input voltage [V]	**31–325**	7.5–**42**	2–**13**	400	15–400	15–450	20–300	60–400
Output voltage [V]	12	3	5	12	3.3–10	10	5	30–60
iVCR range	19	10	5.78	**0**	/	/	/	/
# external comp.	**0**	0	0	0	**2**	1	**7**	**7**
P_{max} [mW]	100	2.1	10	220	300	30	500	29000
Power density [mW/mm^2]	16.4	0.4	1.47	119	0.84[a]	1.8[b]	0.94[a]	/
$\eta_{average}$ [%]	57.3	50.3	60.8	63.6	49.4[c]	**4.2**	67.2	91.8
@ power density [mW/mm^2]	16.4	0.4	0.96	119	0.84[a]	1.8[b]	0.09[a]	/
Max. ripple [%]	4.8	5	/	9.2	5.9	0	3.7	1

[a] Estimated with PCB area
[b] Estimated
[c] Efficiency for a 230 V_{RMS} input

Measurement results demonstrated a 31–325 V DC input range, while achieving a power density of 16.4 mW/mm^2. This outcome advances the maximum input voltage of fully integrated multi-ratio DC-DC converters from 42 V to 325 V, while accomplishing an 11× higher power density at a similar efficiency. Beyond its standalone DC-DC converter applications, it serves as the fundamental building block for the AC-DC converter explored in the subsequent chapter.

Chapter 5
Fully Integrated AC-DC Converter for High Power Density

The preceding chapter introduced a fully integrated DC-DC converter featuring a 31–325 V DC input range.[1] This chapter extends this DC-DC converter into a fully integrated AC-DC converter with a high power density. While the DC-DC converter covers a broad range of mains-rectified AC inputs, it cannot bridge the zero-crossing. Addressing this challenge mandates a sizeable buffer capacitor, imposing constraints on both integration and power density. Consequently, this chapter searches for topology techniques to minimize the size of the buffer capacitor.

This chapter commences by discussing state-of-the-art AC-DC converters and their limitations. It subsequently delves into the challenges entailed in the complete integration of an AC-DC converter and proposes a topology to overcome these obstacles. Section 5.4 details the implementation of this topology, encompassing the control, buffer capacitor, and drivers. To conclude, this chapter demonstrates the measurement results obtained from a prototype and compares them with state-of-the-art approaches.

Parts of this chapter were presented at the Symposium on Very Large Scale Integrated Circuits (VLSI) of 2023 in Kyoto, Japan [84], and have been published in the Journal of Solid-State Circuits (JSSC) in 2023 [86].

5.1 State of the Art and This Work

Figure 5.1 shows the efficiency versus power density for state-of-the-art fully integrated AC-DC converters and this work. Previous integrated solutions [14, 41, 73] employ a capacitive divider since capacitors are suitable for integration. However,

[1] Throughout this book, both peak AC voltages and DC voltages are denoted in volts (V), while AC root mean square (RMS) voltages are specified in volts RMS (V_{RMS}).

© The Author(s), under exclusive license to Springer Nature Switzerland AG 2025
T. Van Daele, F. Tavernier, *Fully Integrated High-Voltage DC-DC and AC-DC Conversion*, Analog Circuits and Signal Processing,
https://doi.org/10.1007/978-3-031-78519-1_5

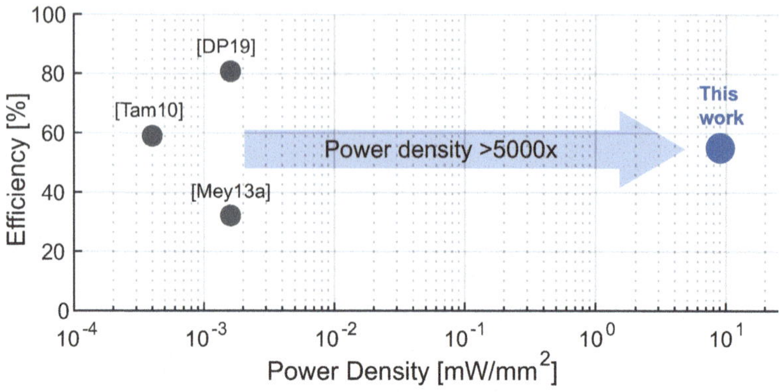

Fig. 5.1 Efficiency at both maximum power and input voltage versus power density for state-of-the-art fully integrated AC-DC converters and this work

due to the low frequency of the mains (50 or 60 Hz) and the limited capacitance density on a chip, they exhibit a high input impedance. This leads to low output powers and, thus, low power densities $\leq 1.6\,\mu W/mm^2$ [14].

In contrast, other previous studies can deliver higher output power, but all depend on external components. External capacitive dividers [54] rely on bulky discrete capacitors. Resistive dividers [35] demand a sizable external buffer capacitor during the AC zero-crossing due to their extremely poor efficiency ($\leq 5\%$ in [35]) and because of the low buffer capacitor voltage at which the buffering starts. Direct coupling [51] needs an even larger capacitor ($15\,\mu F$ in [51]) to store up to 50% of the AC mains, as the input voltage is only passed when it is slightly above the low output voltage. Additionally, they have attained efficiencies of only up to 30% [51]. On the other hand, conventional flyback converters [57] rely on large and expensive transformers and suffer from low efficiencies within the target milliwatt power range [59]. Lastly, resonant/buck converters [60] provide high power and improved efficiency but necessitate at least two external components, one of which is a bulky inductor.

This work proposes the switched-capacitor AC-DC topology, which employs a switched-capacitor DC-DC converter (SCC) to obtain complete integration and high power density. An SCC relies solely on switches and capacitors, allowing for a fully integrated solution in CMOS. Operating the SCC at a high frequency, independent of the mains frequency, delivers significant output power, even with the limited integrated capacitance. Hence, this approach achieves a >5000× higher power density than previous fully integrated designs. Note that this AC-DC converter is compatible with the European 50 Hz 230 V_{RMS} mains like in [35, 41, 60]. This compatibility presents a greater challenge than the 120 V_{RMS} mains as in [14, 51, 54, 73] due to the higher voltage and conversion step required.

5.2 Challenges

While SCCs are a suitable candidate for enhancing the power density of fully integrated AC-DC converters, they face significant challenges. Figure 5.2a shows an AC-DC topology where the rectified AC mains voltage is connected to a wide-input-range SCC without an intermediate buffer capacitor. Figure 5.2b visualizes the corresponding voltage ($V_{in,SCC}$) over time. The SCC can effectively manage the rectified input voltage as long as it exceeds its minimum input voltage ($V_{min,SCC}$). However, when the voltage drops below $V_{min,SCC}$ around the zero-crossing, the SCC becomes incapable of generating the required output voltage. This results in an undesirable output voltage drop.

The conventional solution adds a buffer capacitor (C_{buf}) at the input of the SCC, as illustrated in Fig. 5.3a, with the SCC input voltage plotted in Fig. 5.3b. The buffer capacitor is designed to ensure that the SCC input voltage never drops below $V_{min,SCC}$. However, in this approach, the buffer capacitor starts discharging as soon as it has reached its peak value, well before the actual zero-crossing interval. Consequently, to buffer this long time interval (>5 ms), a substantial buffer capacitor is required. For example, employing the 31–325 V SCC from the previous chapter to deliver 100 mW would require a capacitance of approximately 17 nF rated at 325 V.

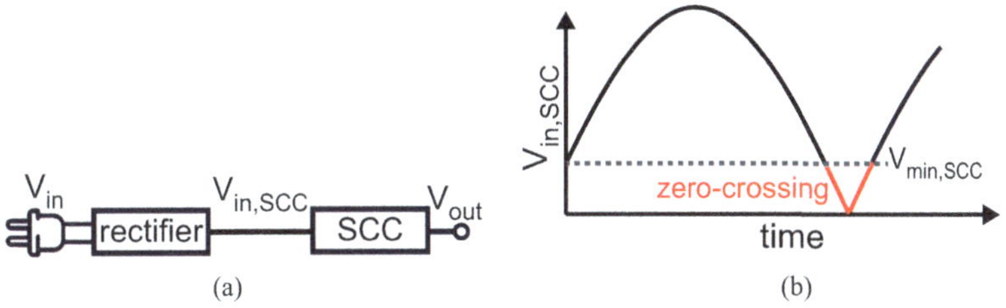

Fig. 5.2 A wide-input-range SCC down-converting the rectified AC input voltage without a buffer capacitor. (**a**) Schematic. (**b**) SCC input voltage over time

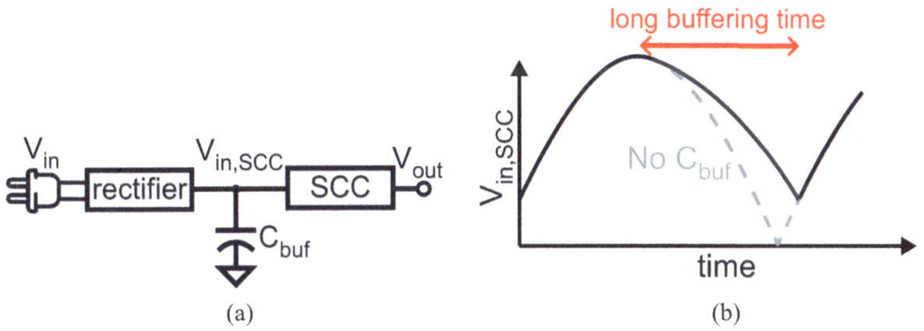

Fig. 5.3 Buffer capacitor directly connected to the SCC input. (**a**) Schematic. (**b**) SCC input voltage over time

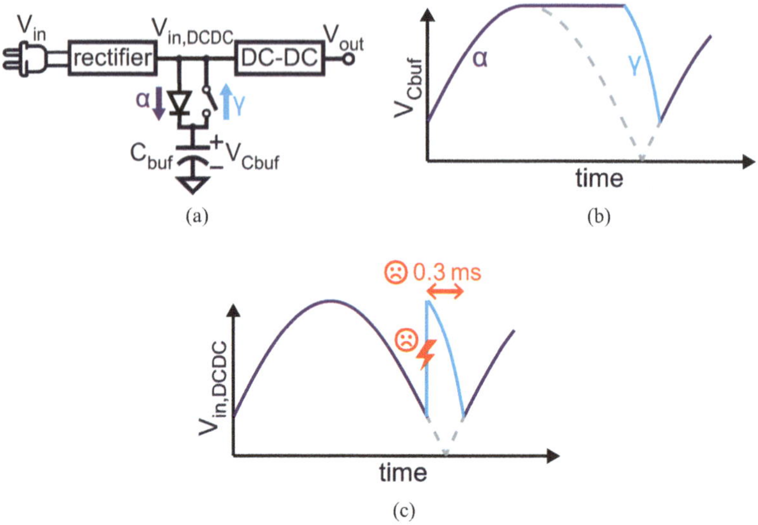

Fig. 5.4 Buffer capacitor connected via a diode and a switch. (**a**) Schematic. (**b**) Buffer capacitor voltage over time. (**c**) DC-DC input voltage over time

Recent work [60] reduced the size of the buffer capacitor when operating a DC-DC converter, as shown in Fig. 5.4a. Rather than directly connecting the buffer capacitor to the input of the DC-DC converter, this approach employs a diode and a switch. In mode α, as depicted in Fig. 5.4b, the rectifier sources power to the DC-DC converter, while a diode simultaneously charges the buffer capacitor to its peak value. Subsequently, when the rectified input voltage decreases, the buffer capacitor maintains the peak voltage because of the unidirectional operation of the diode. When the zero-crossing occurs in mode γ, the buffer capacitor is connected to the DC-DC input via the switch, replacing the rectifier as the input power source for the DC-DC converter. As the buffer capacitor only discharges during the zero-crossing interval, it reduces the buffering time, resulting in the integration of the buffer capacitor.

However, the DC-DC conversion of this technique relies on an external inductor and capacitor. Using an SCC for conversion in this technique enables full integration of the AC-DC converter, yet it poses two key challenges. These challenges are shown in Fig. 5.4c, which depicts the DC-DC input voltage ($V_{in,DCDC}$) over time. Firstly, transitioning from mode α to γ causes a steep voltage rise at the SCC input, surging from 31 V to 325 V in this design. This abrupt increase would break down the low-voltage components in the SCC, as there is no inductor limiting the current. Secondly, the buffering time is long (0.3 ms) due to the high minimum input voltage of the SCC, requiring a substantial chip area for the buffer capacitor.

5.3 Topology

Fully integrating the AC-DC converter requires a considerable buffer capacitor and leads to breakdown in the SCC. To address these issues, this work reduces the buffer capacitor size by introducing a linear regulator, as explained in Sect. 5.3.1. Subsequently, a pre-charging scheme enables a fully integrated SCC in the AC-DC topology while avoiding breakdown. Lastly, Sect. 5.3.3 discusses a linear regulator topology optimal for the AC-DC converter.

5.3.1 Buffering Time Reduction

Figure 5.5 shows an overview of the AC-DC system. All the diodes and switches shown, along with the capacitor, are rated for the maximum peak input voltage (325 V), except for $S_{\alpha,\gamma}$, which is rated for 10 V. The incoming mains voltage is rectified (V_{rect}), which is then converted into the desired DC output voltage (V_{out}). The conversion occurs through three modes (α, β, γ), which are triggered based on the level of V_{rect}. In mode α, when $V_{rect} > 31$ V, a multi-ratio SCC converts V_{rect}. Simultaneously, the buffer capacitor (C_{buf}) is charged to the peak AC voltage. However, as V_{rect} falls below 31 V, the SCC becomes unable of providing V_{out}. Hence, the system transitions to mode β, in which a linear regulator takes over the conversion. The linear regulator operates efficiently because of the low conversion step in this mode. Once the linear regulator input voltage reaches its minimum value (i.e., $V_{rect} = 14$ V), mode β shifts to mode γ. In mode γ, C_{buf} serves as the input power source replacing the mains input, and the SCC converts its stored energy into V_{out}. The peak AC voltage (325 V) is stored during mode α to shorten the buffering

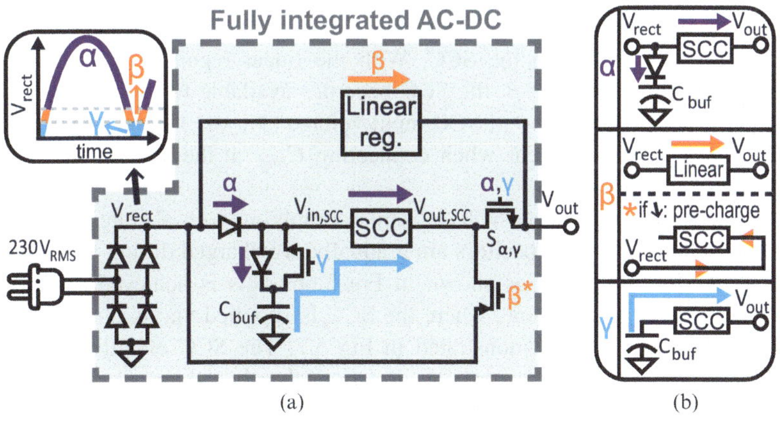

Fig. 5.5 (**a**) Fully integrated AC-DC system overview, (**b**) with the equivalent circuit in every mode

time, thereby decreasing the necessary buffer capacitance, as discussed previously in Sect. 5.2.

The addition of mode β, which uses the linear regulator, substantially decreases the buffer capacitance. Its capacitance depends on the buffering duration (t_{buf}), during which the SCC discharges C_{buf} from the peak AC voltage (V_{peak}) to the minimum SCC input voltage ($V_{min,SCC}$) with efficiency η_{SCC} to maintain the output power (P_{out}):

$$C_{buf} = \frac{2 \cdot P_{out} \cdot t_{buf}}{(V_{peak}^2 - V_{min,SCC}^2) \cdot \eta_{SCC}}. \tag{5.1}$$

The buffering time is restricted by the minimum input voltage (V_{min}) of the DC-DC conversion stage and can be calculated as follows:

$$t_{buf} = \frac{\arcsin\left(\frac{V_{min}}{V_{peak}}\right)}{\pi \cdot f_{AC}} \approx \frac{V_{min}}{\pi \cdot V_{peak} \cdot f_{AC}}, \tag{5.2}$$

where f_{AC} denotes the mains frequency. In the absence of the linear regulator, the minimum input voltage V_{min} in (5.2) corresponds to the high minimum input voltage of the SCC. This scenario leads to an extended buffering time (5.2), and, thus, a sizable buffer capacitor (5.1). However, with the introduction of the linear regulator, the minimum input voltage is reduced from 31 V to 14 V. This leads to a buffer capacitance decrease of approximately 2× (5.2), reducing the overall converter area by 33%.

5.3.2 Pre-charging

The introduction of mode β not only diminishes the buffer capacitance but also prevents breakdown within the SCC. With the linear regulator taking over the voltage conversion in mode β, the SCC becomes available for pre-charging. Pre-charging avoids a steep jump in SCC input voltage ($V_{in,SCC}$) from 31 V to 325 V, which would otherwise occur when connecting C_{buf} at the start of mode γ, as illustrated in Fig. 5.6a.

To increase the SCC input voltage while complying with all rated component voltages, the SCC flying capacitors are gradually pre-charged during the transition from mode β to mode γ, as shown in Fig. 5.6b. This is achieved through the proposed pre-charging scheme, where the SCC is supplied via its output with the low voltage of V_{rect}, as demonstrated in Fig 5.7. The SCC effectively operates as a step-up converter in the opposite direction, thereby taking advantage of the bidirectional nature of SCCs. For this purpose, the SCC is configured to its highest ideal voltage conversion step (N_{max}) and continuously switches. Each switching cycle transfers a restricted charge to the flying capacitors, thereby guaranteeing a

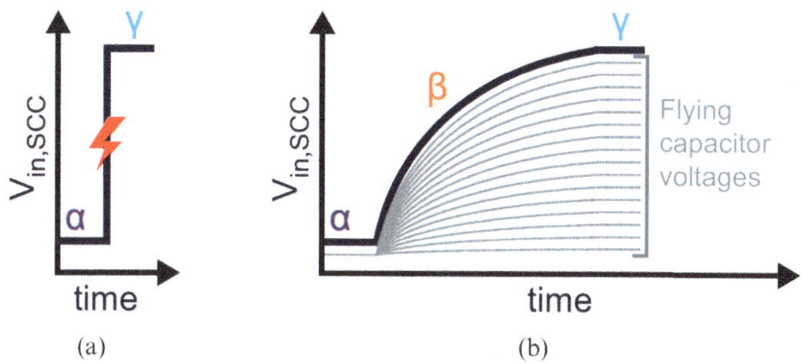

Fig. 5.6 SCC input voltage versus time (**a**) without and (**b**) with pre-charging in mode β. The gray graphs show the gradual pre-charging of the SCC flying capacitors

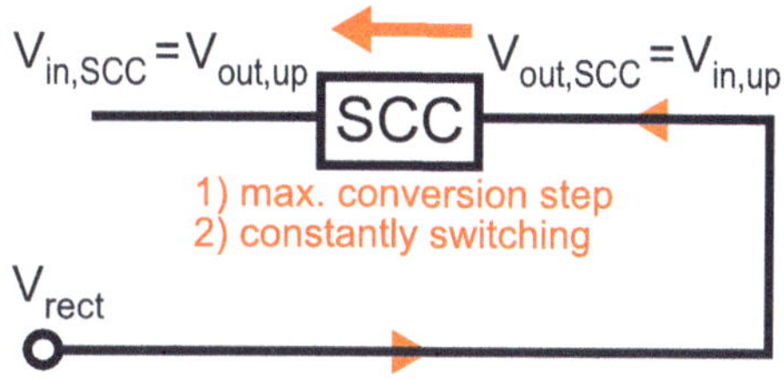

Fig. 5.7 Pre-charging concept

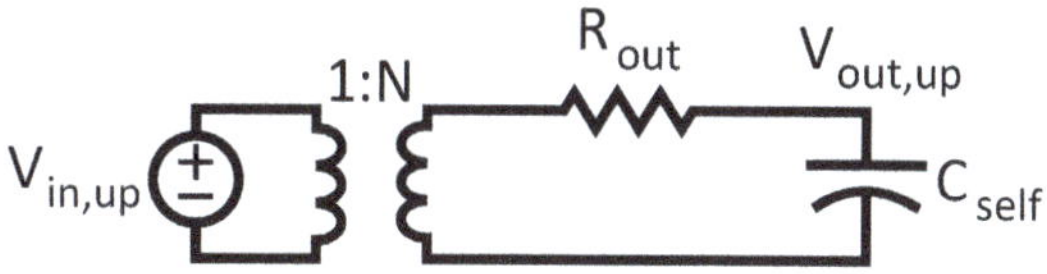

Fig. 5.8 Dynamic equivalent model of an SCC during pre-charging

smooth and controlled buildup of the internal voltages. The charge balance within SCCs ensures that these internal voltages converge to their designated levels. In the remainder of this section, the nodes $V_{in,SCC}$ and $V_{out,SCC}$ of Fig. 5.5 are renamed as $V_{out,up}$ and $V_{in,up}$, respectively, for the sake of clarity in the context of the step-up topology.

The pre-charging should be completed prior to transitioning to mode γ. The rise time of $V_{out,up}$ (=$V_{in,SCC}$) is determined using the dynamic equivalent model of a charge pump [75], as illustrated in Fig. 5.8, which is also valid for SCCs [3]. In this model, the input voltage $V_{in,up}$ is up-converted by a DC transformer, followed by a resistive output impedance (R_{out}). The output impedance remains the same as in the steady-state model of the SCC in Sect. 2.2.2, encompassing the SSL and FSL asymptotes. The steady-state assumption holds because the rise time is substantially longer than the switching period of the converter during pre-charging. Continuing through the model, an equivalent self-capacitance (C_{self}) accounts for the internal flying capacitors' charge time. Assuming a capacitance C_{fly} for all flying capacitors within the $1 : N_{max}$ Dickson step-up topology, the expression for C_{self} is given by

$$C_{self} = \begin{cases} \frac{4N_{max}^2+3N_{max}+2}{12(N_{max}+1)} \cdot C_{fly}, & \text{for even } N_{max}, \\ \frac{4N_{max}^2-N_{max}-3}{12(N_{max}+1)} \cdot C_{fly}, & \text{for odd } N_{max}, \end{cases} \tag{5.3}$$

$$\approx \frac{C_{fly}}{3}. \tag{5.4}$$

To calculate the rise time, first the step-up output voltage ($V_{out,up}$) in function of time (t) is given for charging the self-capacitor with an initial capacitor voltage $V_{out,up}(0)$:

$$V_{out,up}(t) = V_{in,up} - \left(V_{in,up} - V_{out,up}(0)\right) \cdot \exp\left(\frac{-t}{R_{out} \cdot C_{self}}\right), \tag{5.5}$$

with $R_{out} \cdot C_{self}$ representing the time constant. Subsequently, by expressing the time in function of the other variables and by substituting $V_{out,up}(t)$ with 325 V and $V_{out,up}(0)$ with 31 V, the rise time (t_{rise}) is obtained:

$$t_{rise} = R_{out} \cdot C_{self} \cdot \ln\left(\frac{V_{in,up} - 31}{V_{in,up} - 325}\right). \tag{5.6}$$

The design ensures that $V_{in,SCC}$ rises until 325 V within mode β while limiting the losses. With the majority of design variables already selected for efficient down-conversion of the SCC in mode α, only two design variables need to be determined. These are the switching frequency and $V_{in,up}$. The latter is influenced by the timing when $V_{in,up}$ is supplied by V_{rect}, as V_{rect} varies from 31 V to 14 V within mode β. Both design variables have an impact on the rise time, as well as on the power losses. A higher switching frequency reduces the output impedance and thus the rise time (5.6) while operating in the SSL region. However, this comes at the cost of increased parasitic switching losses in the SCC. Conversely, a higher $V_{in,up}$ also leads to a shorter rise time (5.6), but it raises the voltage drop over the output impedance, consequently elevating the output impedance losses. Both the frequency and $V_{in,up}$ are selected to complete pre-charging before mode β ends while approximately balancing the output impedance and switching losses. This balance is not critical as the pre-charging losses are inherently restricted because they occur only once every mains period. Operating the SCC at 170 MHz during the 20.5–18 V interval of V_{rect} corresponds to the optimal clock frequency of the SCC in mode α, while facilitating a reasonable compromise between output impedance and switching losses. This results in simulated pre-charging losses of 399 μW.

5.3.3 Linear Regulator Topology

The topology of the introduced linear regulator is optimized for the AC-DC conversion. It uses a lower-bound clocked hysteretic control [7], as shown in

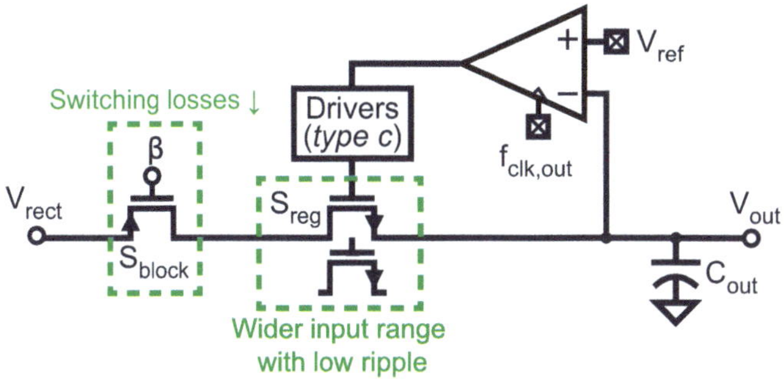

Fig. 5.9 Linear regulator topology with optimizations in green

Fig. 5.9. This control activates the pass transistor (S_{reg}) for one clock period when the output voltage is below the reference voltage (V_{ref}) at the clock edge ($f_{clk,out}$). Throughout this activation period, a current flows to the output, with its magnitude following from the on-resistances of both S_{block} and S_{reg}, as well as the voltage across them. The hysteretic linear regulator provides several benefits in the new context of an AC-DC converter. It can instantaneously alter the effective switching frequency, enabling a rapid transient response. This dynamic behavior is comparable to the SCC, resulting in an excellent output load response throughout the entire AC cycle. Additionally, the hysteretic linear regulator facilitates efficient shutdown of the linear regulator outside mode β. As the linear regulator is not delivering output power in this case, the pass transistor and its control are not switching, thus eliminating their switching losses. The remaining power consumption of the comparator is avoided through repurposing the comparator by the SCC when the linear regulator is off (cf. later in Sect. 5.4.1.2). Lastly, by sharing the output decoupling (C_{out}) with the SCC, the area overhead of the linear regulator is reduced to only 2.8% of the overall AC-DC converter area.

The linear regulator is optimized to handle the 14–31 V input range in mode β while blocking 325 V in mode α. To achieve this, Fig. 5.9 illustrates two techniques that improve the switching losses and the output-voltage ripple, respectively. The first technique splits the pass transistor into two cascaded switches to minimize the switching losses. Specifically, switch S_{block} blocks the high input voltage in mode α, while remaining active throughout the regulating mode β without engaging in any switching. Blocking the high voltage by S_{block} (rated for 325 V) in mode α allows for a low-voltage implementation of the regulating switch (S_{reg}). As a result, S_{reg} is rated for 20 V instead of 325 V, benefiting from a superior on-resistance versus parasitic capacitance trade-off for switching in mode β. Note that the cascaded configuration introduces the additional series resistance of S_{block}. However, this resistance can be mitigated by increasing the width of S_{block}, while its driving losses remain minimal due to its infrequent activation at 200 Hz. Its on-resistance is set at 109 Ω to balance the losses and the area overhead.

The second technique enables a wide input range while maintaining a low output-voltage ripple. The maximum output-voltage ripple (ΔV_{out}) when turning on the pass transistor during an on-time t_{on} while conducting an on-current I_{on} can be expressed as

$$\Delta V_{out} = \frac{I_{on} \cdot t_{on}}{C_{out}}. \tag{5.7}$$

When the linear regulator converts its input voltage V_{rect} into the desired V_{out}, the equation becomes

$$\Delta V_{out} = \frac{(V_{rect} - V_{out}) \cdot t_{on}}{R_{on} \cdot C_{out}}, \tag{5.8}$$

with R_{on} the on-resistance of the pass transistor. Equation (5.8) reveals that the ripple would increase as the input V_{rect} rises. To counteract this growing output-voltage ripple, the equivalent on-resistance is adjusted by modifying the pass transistor width based on the input voltage. The width modification is similar to [7]. However, in this work, it facilitates a wide input range instead of compensating for process, voltage, and temperature (PVT) variations. At higher input voltages, the pass transistor enters the saturation region, inherently increasing its on-resistance, which makes further width adjustments unnecessary. By implementing two resistor values (each 69 Ω in the linear region), the measured output-voltage ripple of the linear regulator is decreased by 33% over the 14–31 V input range (Fig. 5.10), thereby no more dominating in the total AC-DC steady-state ripple.

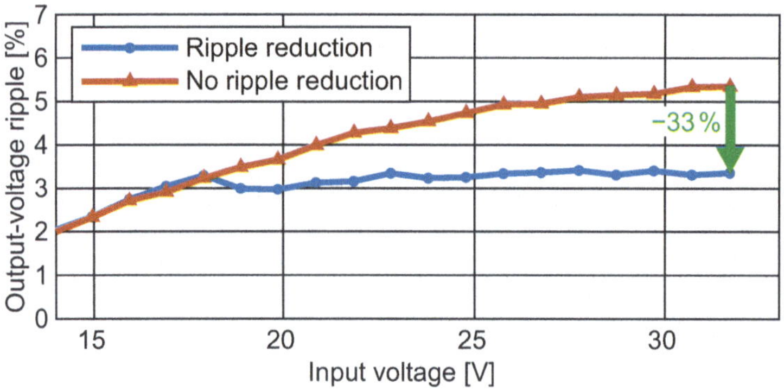

Fig. 5.10 Measured output-voltage ripple versus input voltage for the linear regulator operating at its maximum output power of 100 mW. Comparison of the ripple with and without the implemented ripple reduction technique

5.4 Implementation

To implement the proposed AC-DC converter, this section first delves into the control circuitry. Subsequently, the area of the buffer capacitor is optimized given its substantial influence on the overall area. Afterward, the implementation of the drivers is discussed, which optimizes their area and losses. Finally, the start-up process is discussed.

5.4.1 *Control*

Figure 5.11 illustrates the input and output control within the AC-DC converter. The input control senses the rectified input voltage (V_{rect}) and the buffer capacitor voltage (V_{buf}). It uses this information to determine the operating mode (α, β, γ) and the topology of both the linear regulator and the SCC. On the other hand, the output control monitors the output voltage to adjust the switching frequency of the linear regulator and the SCC. This dynamic regulation maintains a stable output voltage, also in the presence of load and line transients. This subsection discusses the implementation details of both the input and output control.

5.4.1.1 Input Control

Figure 5.12 depicts the input control implementation, which includes seven state bits. The first five bits ($state_{in}$) are derived from the rectified input voltage and the buffer capacitor voltage. To obtain $state_{in}$, both input voltages are first divided by

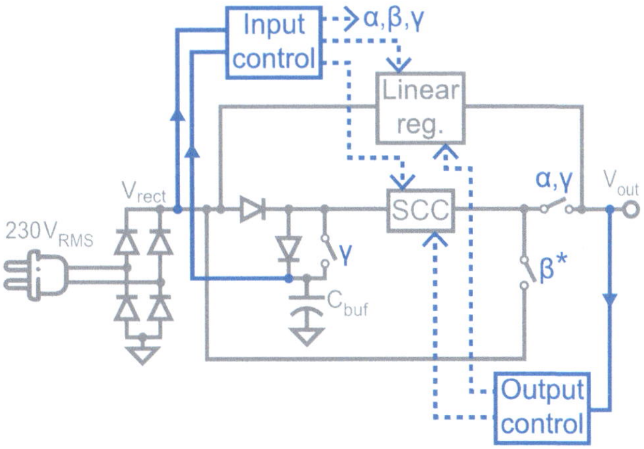

Fig. 5.11 Input control and output control (blue) within the AC-DC converter

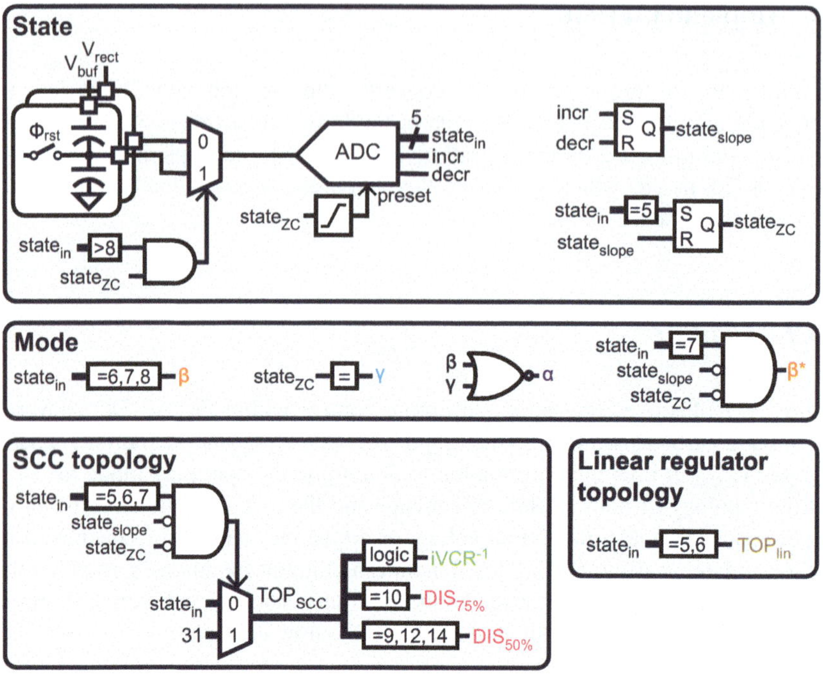

Fig. 5.12 Input control implementation. The colors match those depicted in Fig. 5.13

the resettable capacitive divider introduced earlier in Sect. 4.5.3.1. Subsequently, a multiplexer selects the voltage serving as the input power source. Specifically, the divided V_{buf} is chosen around the zero-crossing, while the divided V_{rect} during the remaining time. Following this, the specialized ADC of Sect. 4.5.3.2 provides $state_{in}$, which is preset at the start of a zero-crossing event to initiate pre-charging. The remaining two state bits are $state_{slope}$ and $state_{ZC}$. The former is assigned a value of 1 when $state_{in}$ has recently been incremented (incr), while the latter is active during zero-crossing events.

The state bits determine the operating mode, as visualized in Fig. 5.13. Mode β (orange) activates the linear regulator when the selected input voltage is below 31 V, corresponding to $state_{in} = 6, 7, 8$. Mode γ (blue) aligns with the value of $state_{ZC}$, and mode α (purple) is active when both mode β and mode γ are off. Within mode β^* (orange), the SCC is pre-charged right before a zero-crossing event. This pre-charging operation occurs when the selected input voltage is decreasing from 20.5 V to 18 V, given that a zero-crossing event is not already happening (i.e., $state_{ZC} = 0$).

Furthermore, these state bits also configure both the SCC and the linear regulator topology. In the case of the SCC, the SCC topology bits (TOP_{SCC}) map to the iVCR (green) and to the disable signals $DIS_{50\%}$ and $DIS_{75\%}$ (pink) that control the deactivation of 50% and 75% of the stages, respectively. TOP_{SCC} follows from

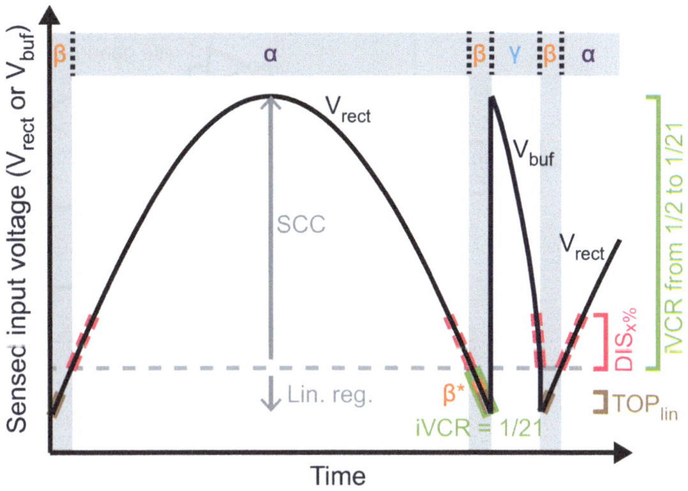

Fig. 5.13 Conceptual input control annotated on the selected input voltage (i.e., V_{rect} or V_{buf}) over time

the selected input voltage via $state_{in}$, except right before a zero-crossing event. TOP_{SCC} becomes 31 prior to a zero-crossing, corresponding to the maximum conversion step (i.e., 21) set for pre-charging. As for the linear regulator, its topology enables an additional parallel pass transistor via the control signal TOP_{lin} (brown). The value of this control signal is determined by $state_{in}$ to optimize the output-voltage ripple (cf. Sect. 5.3.3).

5.4.1.2 Output Control

Figure 5.14 shows the output control using a lower-bound clocked hysteretic control, which generates the control signals for both the SCC ($\Phi_{0°}$, $\Phi_{90°}$, $\Phi_{180°}$, $\Phi_{270°}$) and the linear regulator (Φ_{lin}). The output control starts by dividing the output voltage within the comparator supply range. Following this, the comparator evaluates whether the divided output voltage has dropped below the reference voltage, initiating charge events accordingly. The pulse skipper then skips every second consecutive charge event, as was introduced in Sect. 3.5.4. The described control circuitry is shared between the SCC and the linear regulator, enabling a smooth transition between the two without introducing overshoots or voltage droops. Furthermore, this sharing arrangement results in reduced quiescent power consumption, as it necessitates only a single clocked comparator.

Following this shared segment, the control paths diverge. The value of mode β determines whether the charge event is propagated to the SCC or the linear regulator. In the case of the linear regulator, a pulse generator triggers its pass gate, turning it on for 9 ns. Conversely, when the SCC is selected, the control circuitry based

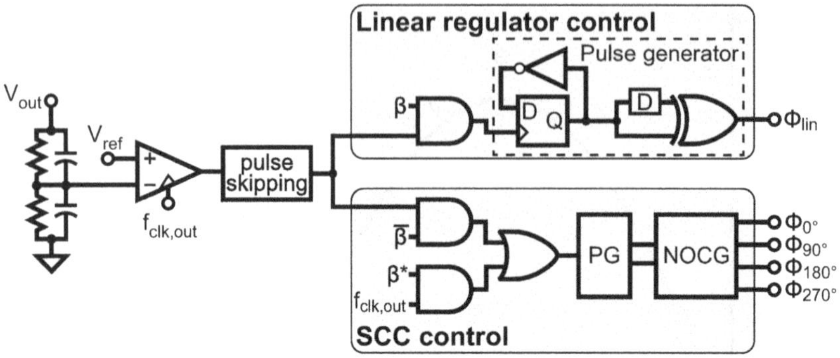

Fig. 5.14 Output control implementation

on [64] creates the SCC phases. The phase generator (PG) generates interleaved signals, and the nonoverlapping clock generator (NOCG) prevents shoot-through in the SCC. Furthermore, during pre-charging ($\beta^* = 1$), the SCC is consistently triggered at the output control clock frequency.

5.4.2 Buffer Capacitor Implementation

As the buffer capacitor (C_{buf}) requires a large capacitance, an effort was made to minimize its area by optimizing its capacitance density. Its bottom plate is fixed to the ground in the AC-DC topology (Fig. 5.5a). By selecting the p-well plate in Fig. 5.15a as the top plate (TP), the parasitic BOX capacitor (C_{BOX}) resides between the top plate and the ground like the original capacitors between the metal plates and the p-well. This addition of the BOX capacitor in parallel with the original buffer capacitor augments the overall buffer capacitance. Considering x capacitor plates initially (comprising $x - 1$ metal plates and 1 p-well plate) and assuming the BOX thickness matches the intermetal thicknesses, the gain in capacitance density is as follows:

$$gain_{\rho} = \frac{x+1}{x}. \tag{5.9}$$

In this design, which originally features six metal layers, this would lead to a capacitance density increase of 14%.

The implementation of this technique in a prototype did not yield the desired result. When rapidly charging the BOX capacitor to a high voltage, the measured capacitance density gain proves to be negligible. The time available for minority charge carriers to form an inversion layer is insufficient, resembling the behavior of a MOS capacitor. Instead, as depicted in Fig. 5.15b, the p-substrate goes into deep depletion, introducing a capacitor C_{depl} in series with the BOX capacitor. This series

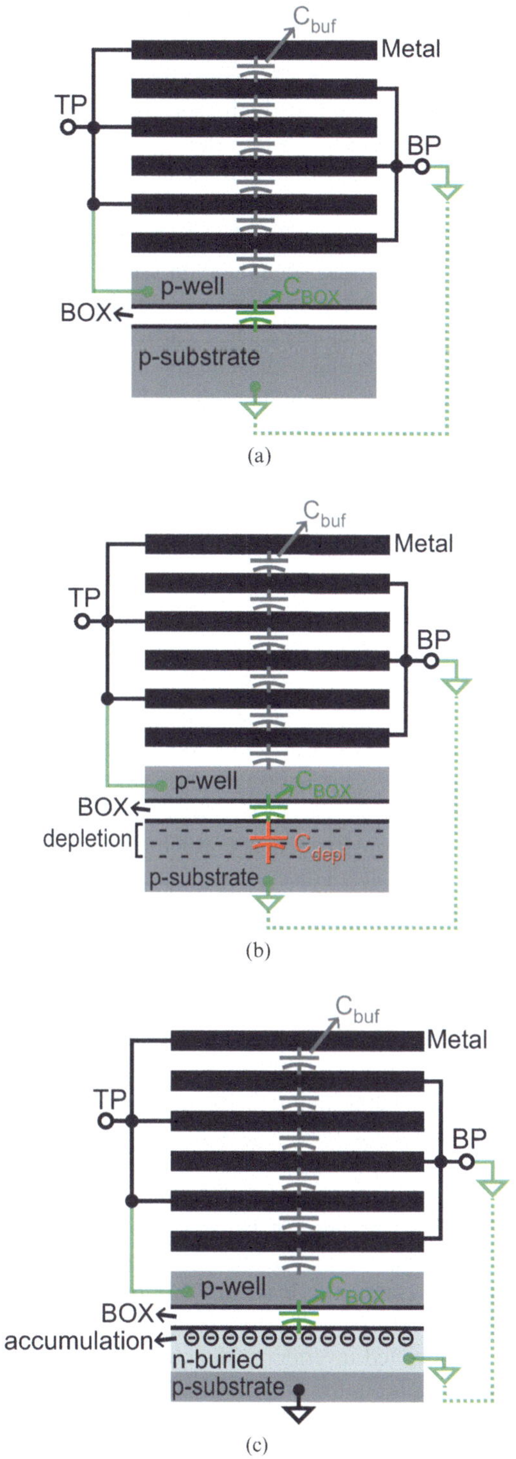

Fig. 5.15 Buffer capacitor implementation. (**a**) Placing the BOX capacitor in parallel with the original buffer capacitor. (**b**) The p-substrate goes into deep depletion at high frequencies. (**c**) Inserting an n-buried layer leads to accumulation instead

capacitor is small due to the high voltage across the BOX, rendering the overall gain negligible. Despite the buffer capacitor voltage increasing at a rate of only $\leq 0.1\,\text{V}/\mu\text{s}$ in the AC-DC topology, measurements revealed that the additional BOX capacitance remains insignificant, showing the persistence of deep depletion in the p-substrate. As such, 50 Hz should still be considered a frequency too high for the minority charge carriers to form an inversion layer.

To address this issue in future designs, an n-buried layer should be inserted below the BOX, as illustrated in Fig. 5.15c. Since the n-buried layer is n-type rather than p-type, it goes into accumulation rather than depletion when a positive voltage is applied to the top plate. The accumulation layer can form even at high frequencies, facilitated by majority charge carriers. This approach avoids the series depletion capacitor associated with an inversion layer. An additional advantage of the n-buried layer is that the current flows through this layer to the ground connection rather than traversing the substrate, mitigating substrate noise. Furthermore, the area overhead to introduce the n-buried layer is limited. The n-buried layer is grounded, while the BOX blocks the high voltage, resulting in no voltage drop across the handle-wafer diode. This obviates the need for extensive high-voltage termination structures.

5.4.3 Drivers

The driver implementation is specialized for the specific switch it operates. The drivers in the SCC were covered in Sect. 4.5.2.

The mode switches share the same driver characteristics as the configuration switches within the SCC. Their characteristics involve the operation of large switches at low frequencies, which require identical driving circuits for both switch types. An in-depth discussion of these level-shifters and charge pumps was given in Sect. 4.5.2 for the configuration switches, which also applies to the mode switches.

The regulating switch of the linear regulator requires another charge delivery characteristic (driver *type c* in Fig. 5.9), as depicted in Fig. 5.16a. It requires a charge burst during its 3% active phase within a mains period while compensating for leakage during the remaining idle phase with a smaller charge. To minimize the losses, the driver combines the level-shifter (LS) from [15] with the previously proposed charge pump (CP) of Sect. 4.5.2, as illustrated in Fig. 5.16b. Both the level-shifter and the charge pump deliver charge to the high-voltage domain, albeit in different operating modes and with different bootstrap capacitances. During the active phase, the level-shifter frequently switches its large bootstrap capacitors (C_{LS}, 4 pF) because of the hysteretic control, which enables the delivery of the burst charge. During the off-time, the level-shifter remains idle, leaving the charge pump to counteract leakage. This hybrid charge delivery scheme minimizes the parasitic losses by switching only the smaller bootstrap capacitors (C_{CP}, 150 fF) and associated inverters at the charge pump frequency during the long idle phase. Furthermore, in the burst mode, the level-shifter only switches (f_{lin}) when charge delivery is required thanks to the hysteretic control, scaling its switching losses

(a)

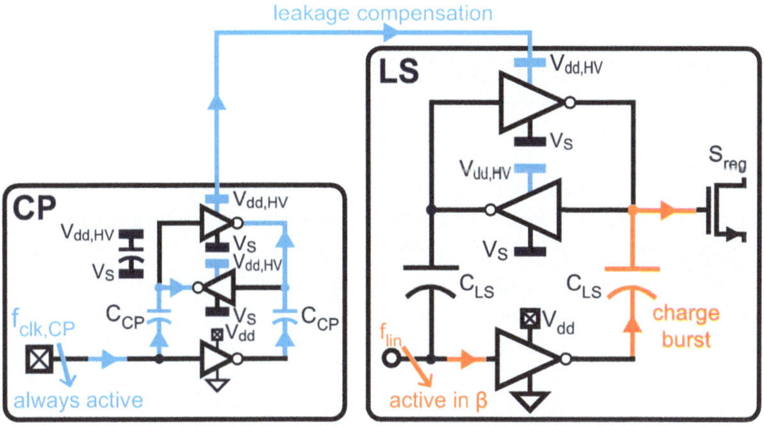

(b)

Fig. 5.16 Driver of the regulating switch of the linear regulator. (**a**) Charge (q) delivery characteristic. (**b**) Schematic

with the output power. Note that the area impact of this driver implementation is negligible, given that the linear regulator contains only two regulating switches to drive.

In summary, Fig. 5.17 presents a high-level overview of all drivers utilized in the AC-DC converter. For the configuration and mode switches, the proposed charge pump delivers the charge, enabling a small level-shifter. The charge pump operates at a frequency 7500× higher than that of the level-shifter, resulting in a small size as well. In the case of the power switches, the charge is delivered by a level-shifter, with the charge pump being omitted. This approach reduces the need for high-voltage isolation, which minimizes the area of each of the small drivers. Lastly, for the linear regulator, the large level-shifter manages the charge bursts during the brief active phase, whereas the smaller charge pump counteracts leakage throughout the extended idle phase.

5.4.4 Start-up

The start-up sequence of the AC-DC converter closely resembles that of the DC-DC converter, as detailed in Sect. 3.5.5. In summary, this procedure begins with the elevation of the external control voltages, followed by setting its latches to their

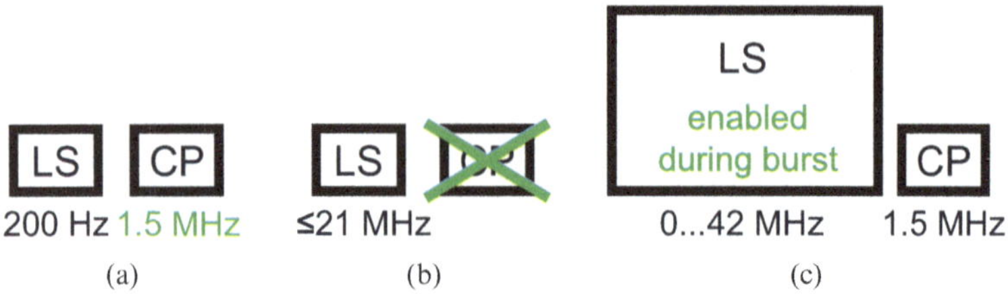

Fig. 5.17 High-level overview of charge delivery by level-shifter (LS) and/or charge pump (CP) for the (**a**) configuration and mode switches, (**b**) power switches, and (**c**) linear regulator switches

designated state. Subsequently, the input voltage commences the 50 Hz AC cycle, starting at 0 V. Note that the converter does not support hot plugging, as it requires an initial input voltage of 0 V for start-up.

Nevertheless, hot plugging can be facilitated with minimal power and area overhead. In particular, a high-voltage switch can be placed immediately after the rectifier to initially block high voltage from reaching the SCC, thereby preventing breakdown. Concurrently, a shunt regulator can generate the supply rails from the rectified input voltage. Once these supplies are established and the AC input crosses 0 V, the series input switch can be safely closed. As the rectified AC input slowly rises afterward, the SCC is continuously switched, effectively pre-charging the flying capacitors.

Implementing hot plugging requires two high-voltage switches: one serving as the series switch after the rectifier and another within the shunt regulator. Importantly, their losses during steady-state operation are minimal. In steady state, the shunt regulator is deactivated, as the supplies can be efficiently generated from the SCC output. Furthermore, the conduction losses of the series input switch can be mitigated by enlarging its size, while the switching losses remain negligible because of its continuous on-state.

5.5 Chip Measurements

A prototype that embodies the proposed topology and implementation techniques is fabricated. This prototype converts 230 V_{RMS} into a stable 12 V DC output voltage while delivering a peak output power of 100 mW. The design is fully integrated into a 180 nm high-voltage CMOS SOI technology. The micrograph in Fig. 5.18 reveals an active area of 11.1 mm^2. This area encompasses the buffer capacitor (1 nF) and the output decoupling (0.5 nF), which occupy 37% and 6.5% of the active area, respectively. The buffer capacitor is rated for 325 V and implemented as a sandwich capacitor (horizontal parallel plates), while the output decoupling capacitor is rated for 14 V and is a finger capacitor (vertical parallel plate).

Even though the supply voltage is provided externally for measurement flexibility, it could be generated on-chip (cf. Sect. 3.6.1). During start-up, a shunt regulator

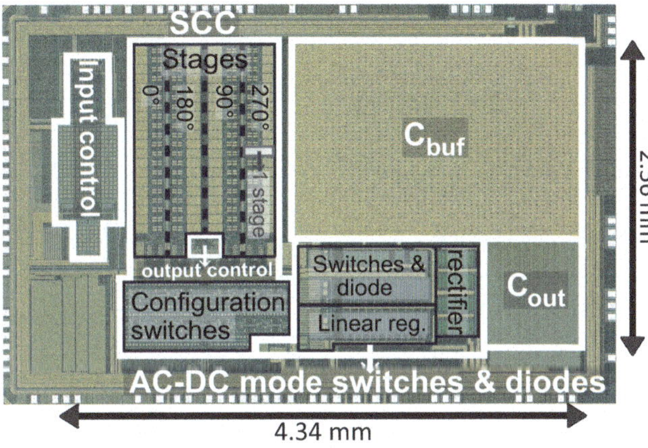

Fig. 5.18 Die micrograph of the fully integrated AC-DC converter

can temporarily generate this supply from the rectified input voltage. During steady-state operation, the supply can be derived from the converter output.

5.5.1 Measurement Setup

The measurement setup is nearly identical to the one given in Sect. 4.6.1. The only difference is that the input voltage source provides an AC input voltage instead of a variable DC input voltage.

5.5.2 Transients

Figure 5.19 shows two output-voltage transients for a 230 V_{RMS}-to-12 V_{DC} AC-DC conversion. In Fig. 5.19a, an applied load step goes from 2 mW to 100 mW and back with a transition time of 50 ns at the maximum input voltage. The peaks in the output voltage result from SCC topology transitions when changing the iVCR from 1/4 to 1/3 and from 1/3 to 1/2 (cf. Sect. 4.4.2). Figure 5.19b and 5.19c shows close-up views of the output voltage during the applied load steps from 2 mW to 100 mW and back with a transition time of $<$50 ns. The rapid output load-step response demonstrates no overshoot or droop.

Figure 5.20 shows a series of 400 load steps between 2 mW and 100 mW spread over one AC period, preceded by a 2 mW load and followed by a 100 mW load. It shows the output load-step response across the entire range of input voltages, each with its corresponding converter topology. The output response in the load-step region remains confined within the peaks of the adjacent 2 mW and 100 mW

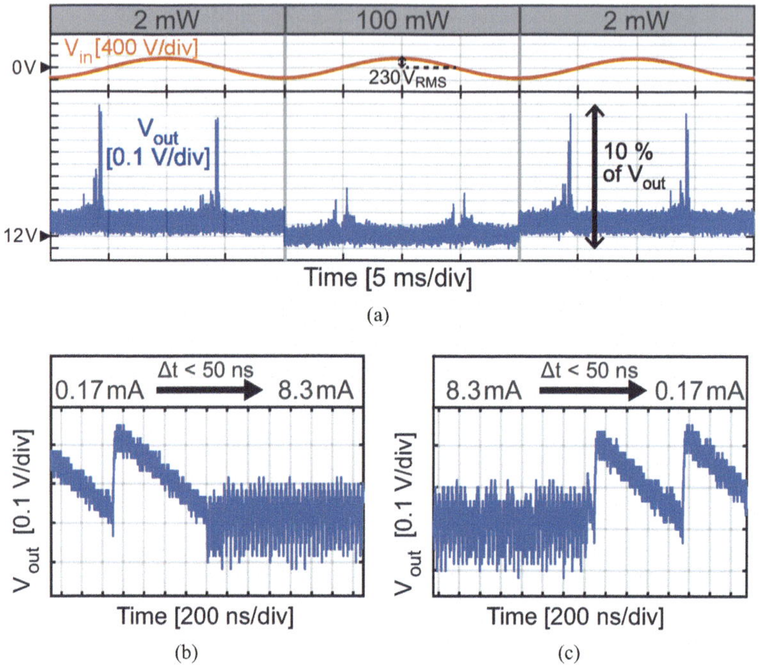

Fig. 5.19 (**a**) Measured transients with $V_{in} = 230\,\text{V}_{\text{RMS}}$ and $V_{ref} = 12\,\text{V}$. It includes two load steps from 2 mW to 100 mW and back with <50 ns transition times. (**b**) and (**c**) Zoomed-in output-voltage responses corresponding to these load steps

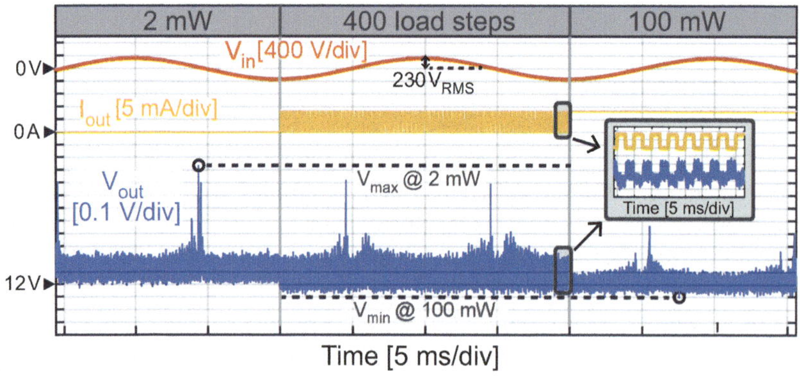

Fig. 5.20 Measured transients with $V_{in} = 230\,\text{V}_{\text{RMS}}$ and $V_{ref} = 12\,\text{V}$. It covers a sequence of 400 load steps along an AC period with <50 ns transition times

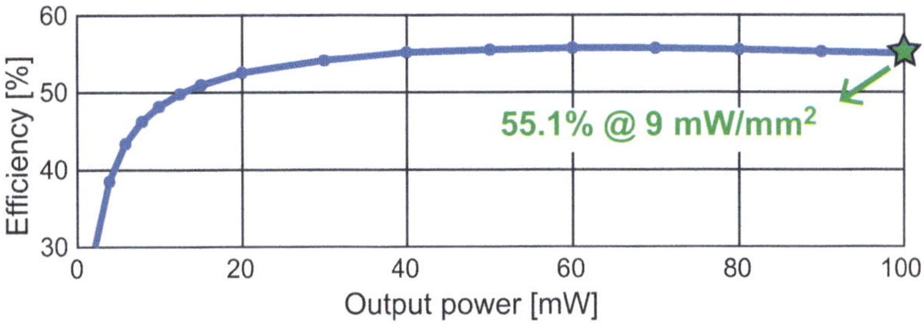

Fig. 5.21 Measured full-system efficiency versus output power for a 230 V_{RMS}-to-12 V_{DC} AC-DC conversion

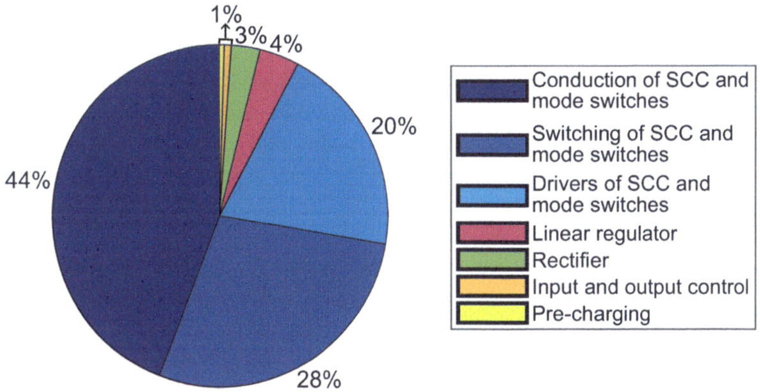

Fig. 5.22 Measured loss breakdown for a 230 V_{RMS}-to-12 V_{DC} AC-DC conversion at 100 mW output power

regions (V_{min}, V_{max}), demonstrating the absence of droop and overshoot during load steps across all converter topologies.

5.5.3 *Efficiency*

The efficiency measurements include all losses, excluding the generation of the external clock and reference bias. Figure 5.21 shows the measured efficiency versus output power for a 230 V_{RMS}-to-12 V_{DC} AC-DC conversion. The AC-DC converter attains an efficiency (η) of 55.1% at a power density of 9 mW/mm^2.

Figure 5.22 illustrates the loss distribution for a 230 V_{RMS}-to-12 V_{DC} conversion operating at the peak output power of 100 mW. The category *Switching of SCC and mode switches* includes the periodic charging of both the drain-source capacitors of the associated switches and the parasitic capacitors of the SCC flying capacitors.

Within the pie chart, the predominant share is attributed to the SCC and the mode switches. The SCC is the main contributor to this share, while the mode switches have minimal impact because of their low on-resistance and low switching frequency of 200 Hz. In the SCC segments, the losses are balanced by selecting the switching frequency (42.5 MHz at maximum output power) and the total power switch on-resistance [8].

5.5.4 *Comparison*

Table 5.1 compares this work with state-of-the-art AC-DC converters. This work does not need any external component contrasting with [35, 51, 54, 60], with [54] requiring as many as seven external components. Furthermore, this work attains higher efficiency than [35, 51] and a similar efficiency to [60] while preventing a bulky inductor. Compared to earlier fully integrated designs [14, 41, 73], it advances the power density by a factor of $>5000\times$. Additionally, it showcases an efficiency superior to [41], while being compatible with 230 V_{RMS} unlike [14, 73] and being regulated in contrast to [14].

5.6 Conclusion

This chapter developed a fully integrated AC-DC converter with a high power density. Prior integrated works relied on capacitive dividers, which suffer from a lower power density. This limitation arises because their input impedance is inherently linked to the low frequency of the AC mains. In contrast, this work introduced an AC-DC converter that incorporates an SCC operating at a significantly higher frequency, independent of the AC mains frequency, enabling higher power densities.

More specifically, this chapter extended the multi-ratio SCC of the preceding chapter into an AC-DC converter. The key challenge was the sizable buffer capacitor required to bridge the zero-crossing. Although recent literature introduced a reduction technique, this approach would result in breakdown and would demand a substantial buffer capacitor area for an SCC. As a solution, this chapter introduced an AC-DC topology that pre-charges the SCC to avoid breakdown. Additionally, the topology introduced a linear regulator to minimize the buffering time, thereby reducing the required buffer capacitor area by half.

The measured power density is 9 mW/mm^2 at 55.1% efficiency. This advances the power density of fully integrated state-of-the-art AC-DC converters by $>5000\times$.

Table 5.1 Comparison with state-of-the-art AC-DC converters

	Fully integrated				Not fully integrated			
	This work	JSSC13 [41]	JSSC19 [14]	ESSCIRC10 [73]	JESTPE18 [35]	JSSC78 [51]	JSSC22 [60]	ISSCC20 [54]
Technology [nm]	180, SOI	350	350	130	350	/,SOS	180,SOI	/
Tech. V_{max} [V]	400	/	/	/	700	/	400	/
Topology	SCC	Capacitive division	Capacitive division	Capacitive division	Resistive division	Direct coupling	Resonant + buck	Capacitive division
AC V_{in} [V_{RMS}]	230	230	**120**	**120**	230	**120**	230	**120**
DC V_{out} [V]	12	3.3	9.5[a]	3.5 – 4	3.3	10[a]	3.3 – 10	3.3
# external comp.	**0**	0	0	0	**1**	**1**	**2**	**7**
P_{max} [mW]	100	0.01	0.015	0.0015	3	100	300	600
Power density [mW/mm^2]	**9.0**	**0.0016**	**0.0016**	**0.0004**	/	/	0.84[b]	/
η @ P_{max} [%]	55.1	32	80.7	59	2 – 3	30	49.4	74
Max. ripple [%]	10.8	5	22.5	5.76	6	1	5.9	0.3

[a] Unregulated
[b] Estimated with PCB area

Chapter 6
Conclusion

6.1 Overview of the Work

High-voltage power sources are ubiquitous, ranging from the AC mains in buildings to the increasing DC bus voltages in the automotive industry and data centers (e.g., 400 V). In contrast, emerging low-power applications, like home automation, LED drivers, and control circuits, demand low supply voltages ($\leq$5 V). Bridging this voltage gap requires a voltage converter handling the high input voltage while operating in the milliwatt power range. Complete integration of such converters enables cost-effective, reliable, and compact solutions. Yet, conventional approaches mandate at least two external components, attributed to their inherent low duty cycles and dependence on an inductor.

This work explored switched-capacitor converters (SCCs) as a potential solution. They solely rely on switches and capacitors and maintain a high duty cycle regardless of the conversion ratio, paving the way for full integration. Before delving into SCCs at high voltage, Chap. 2 elucidated the fundamental concepts. It discussed the operational principle, the loss mechanisms, and the control strategies.

Despite their suitability, prior fully integrated SCCs are confined to input voltages up to 42 V. Chapter 3 elevated this limit to 400 V DC, addressing challenges like diminished component quality and increased parasitic losses. For this, the proposed DC-DC topology leverages the increased dielectric energy storage at high voltage and mitigates the sidewall coupling. It was implemented in a high-voltage CMOS technology, featuring custom high-voltage capacitors and drivers, along with an improved control circuit. The measured prototype advances the maximum input voltage of prior fully integrated DC-DC converters by 9×, while boosting the power density by 270×. It also achieves the largest conversion step and the highest efficiency in class.

Chapter 4 broadened the fixed 400 V DC input voltage of Chap. 3 into a wide 31 V–325 V DC input range. This enhancement not only facilitates applications

© The Author(s), under exclusive license to Springer Nature Switzerland AG 2025
T. Van Daele, F. Tavernier, *Fully Integrated High-Voltage DC-DC and AC-DC Conversion*, Analog Circuits and Signal Processing,
https://doi.org/10.1007/978-3-031-78519-1_6

requiring a variable DC input range, but also lays the foundation for the AC-DC converter discussed in Chap. 5. Accommodating such a wide input range introduces numerous configuration switches in the conduction path, increasing the losses. To mitigate this challenge, several topology techniques were proposed. These include the reconfiguration of DC nodes, gradual parallelization, the adoption of low-voltage configuration switches, and the optimization of the iVCRs. The implementation of this DC-DC topology was enabled through high-voltage input sensing and compact drivers. The resulting prototype elevates the maximum input voltage of previous fully integrated multi-ratio converters from 42 V to 325 V, while accomplishing an 11× higher power density at comparable efficiency.

Chapter 5 extended the multi-ratio DC-DC converter of Chap. 4 into an AC-DC converter, introducing the switched-capacitor AC-DC converter. As the AC input frequency is significantly slower than the SCC switching frequency, the AC input effectively behaves like a DC input. Hence, after a bridge rectifier, the multi-ratio SCC can follow the input voltage, despite its DC-DC nature. Nevertheless, with a minimum input voltage of 31 V, a sizable buffer capacitor is necessary to bridge the zero-crossing. While recent work had minimized the buffer capacitance for inductive converters, this approach would lead to breakdown in an SCC, while still demanding a considerable buffer capacitance. To tackle this challenge, this work introduced an AC-DC topology that reliably pre-charges the SCC. In addition, it incorporated an optimized linear regulator, halving the buffer capacitance. The resulting power density measures $9\,\mathrm{mW/mm^2}$ at 55.1% efficiency, advancing the power density of prior fully integrated AC-DC converters by $>5000\times$.

References

1. Antoniou, M., Udrea, F., Tee, E. K. C., Pilkington, S., Pal, D. K., & Hoelke, A. (2012). Compact three-dimensional silicon termination solutions for high voltage SOI SuperJunction. In *2012 24th International Symposium on Power Semiconductor Devices and ICs* (pp. 89–92).
2. Baliga, B. (1989). Power semiconductor device figure of merit for high-frequency applications. *IEEE Electron Device Letters, 10*(10), 455–457.
3. Ballo, A., Grasso, A. D., Palumbo, G., & Tanzawa, T. (2022). A charge loss aware advanced model of Dickson voltage multipliers. *IEEE Access, 10*, 118082–118092.
4. Bang, S., Blaauw, D., & Sylvester, D. (2016). A successive-approximation switched-capacitor DC–DC converter with resolution of $V_{IN}/2^N$ for a wide range of input and output voltages. *IEEE Journal of Solid-State Circuits, 51*(2), 543–556.
5. Bang, S., Seo, J.-S., Chang, L., Blaauw, D., & Sylvester, D. (2016). A low ripple switched-capacitor voltage regulator using flying capacitance dithering. *IEEE Journal of Solid-State Circuits, 51*(4), 919–929.
6. Barner, A., Wittmann, J., Rosahl, T., & Wicht, B. (2016). A 10 MHz, 48-to-5 V synchronous converter with dead time enabled 125 ps resolution zero-voltage switching. In *2016 IEEE Applied Power Electronics Conference and Exposition (APEC)* (pp. 106–110).
7. Bulzacchelli, J. F., Toprak-Deniz, Z., Rasmus, T. M., Iadanza, J. A., Bucossi, W. L., Kim, S., Blanco, R., Cox, C. E., Chhabra, M., LeBlanc, C. D., Trudeau, C. L., & Friedman, D. J. (2012). Dual-loop system of distributed microregulators with high DC accuracy, load response time below 500 ps, and 85-mV dropout voltage. *IEEE Journal of Solid-State Circuits, 47*(4) (2012), 863–874.
8. Butzen, N., & Steyaert, M. (2020) *Advanced multiphasing switched-capacitor DC-DC converters*. Springer.
9. Butzen, N., & Steyaert, M. S. J. (2016). Scalable parasitic charge redistribution: Design of high-efficiency fully integrated switched-capacitor DC–DC converters. *IEEE Journal of Solid-State Circuits, 51*(12), 2843–2853.
10. Butzen, N., & Steyaert, M. S. J. (2017) Design of soft-charging switched-capacitor DC–DC converters using stage outphasing and multiphase soft-charging. *IEEE Journal of Solid-State Circuits, 52*(12), 3132–3141.
11. Choi, M., Kye, C.-H., Oh, J., Choo, M.-S., & Jeong, D.-K. (2019). 27.7 a synthesizable digital AOT 4-phase buck voltage regulator for digital systems with 0.0054 mm^2 controller and 80 ns recovery time. In *2019 IEEE International Solid-State Circuits Conference - (ISSCC)* (pp. 432–434).

© The Author(s), under exclusive license to Springer Nature Switzerland AG 2025

T. Van Daele, F. Tavernier, *Fully Integrated High-Voltage DC-DC and AC-DC Conversion*, Analog Circuits and Signal Processing,
https://doi.org/10.1007/978-3-031-78519-1

12. Cui. PBO-1 AC-DC power supply datasheet. Retrieved July 12, 2022, from https://www.cui.com/product/resource/pbo-1-b.pdf
13. Cui. VSK-S1-5U AC-DC power supply datasheet. Retrieved July 12, 2022, from https://www.mouser.be/datasheet/2/670/vsk_s1-1313103.pdf
14. De Pelecijn, E., & Steyaert, M. S. J. (2019) A fully integrated switched-capacitor-based AC–DC converter for a 120 VRMS mains interface. *IEEE Journal of Solid-State Circuits, 54*(7), 2009–2018.
15. De Pelecijn, E., & Steyaert, M. S. J. (2020) Stacking isolated SC cores for high-voltage wide input range monolithic DC–DC conversion. *IEEE Journal of Solid-State Circuits, 55*(10), 2639–2648.
16. Du, L., & Ma, D. B. (2020). A 48 V-to-1 V buck-assisted active-clamp forward converter with reduced voltage stress for datacenter applications. In *2020 IEEE Energy Conversion Congress and Exposition (ECCE)* (pp. 5442–5446).
17. El-Damak, D., Bandyopadhyay, S., & Chandrakasan, A. P. (2013) A 93% efficiency reconfigurable switched-capacitor DC-DC converter using on-chip ferroelectric capacitors. In *2013 IEEE International Solid-State Circuits Conference Digest of Technical Papers* (pp. 374–375).
18. Energy5. What voltage do Tesla cars run at? Retrieved January 11, 2024, from https://energy5.com/what-voltage-do-tesla-cars-run-at(why-is-it-so-high)
19. Foray, E., Allard, B., Martin, C., & Frattini, G. (2019) Topologies for high-voltage low-power integrated DC-DC converter. In *2019 21st European Conference on Power Electronics and Applications (EPE '19 ECCE Europe)* (pp. P.1–P.8).
20. Foray, E., Martin, C., & Allard, B. (2020) Design and optimization of high-efficiency 1 W 500 V–12 V isolated low-cost DC/DC converter. In *2020 22nd European Conference on Power Electronics and Applications (EPE'20 ECCE Europe)* (pp. P.1–P.11).
21. Gehl, A., Cherniak, K., Kharko, O., Marin, G., Bacinski, P., Prämaßing, F., & Wicht, B. (2023). A 2-8 V Vin 670 mA scalable multi-ratio SC DCDC converter for MCU integration in 28 nm CMOS achieving 91% peak efficiency. In *ESSCIRC 2023- IEEE 49th European Solid State Circuits Conference (ESSCIRC)* (pp. 425–428).
22. Helix Semiconductors. High voltage MUX capacitor voltage reducer plus programmable output regulator IC. Retrieved August 26, 2021, from https://www.helixsemiconductors.com/uploads/cms_uploads/2017/03/helixsemi-muxcapldo-abstract-1489785236.pdf
23. Hölke, A., Antoniou, M., & Udrea, F. (2020). Partial SOI as a HV platform technology for power integrated circuits. In *2020 32nd International Symposium on Power Semiconductor Devices and ICs (ISPSD)* (pp. 435–438).
24. Jain, R., Kim, S. T., Vaidya, V., Ravichandran, K., Tschanz, J. W., & De, V. Conductance modulation techniques in switched-capacitor DC-DC converter for maximum-efficiency tracking and ripple mitigation in 22 nm tri-gate CMOS. *IEEE Journal of Solid-State Circuits, 50*(8), 1809–1819.
25. Kaufmann, M., Lueders, M., Kaya, C., & Wicht, B. (2020). 18.2 a monolithic e-mode GaN 15 W 400 V offline self-supplied hysteretic buck converter with 95.6% efficiency. In *2020 IEEE International Solid- State Circuits Conference - (ISSCC)* (pp. 288–290).
26. Kiani, M. H., & Stauth, J. T. (2017). Optimization and comparison of hybrid-resonant switched capacitor DC-DC converter topologies. In *2017 IEEE 18th Workshop on Control and Modeling for Power Electronics (COMPEL)* (pp. 1–8).
27. Kudva, S. S., & Harjani, R. (2013). Fully integrated capacitive DC–DC converter with all-digital ripple mitigation technique. *IEEE Journal of Solid-State Circuits, 48*(8), 1910–1920.
28. Lange, A., Heinz, S., Erler, K., Ebest, G., Lerner, R., Eckoldt, U., & Schottmann, K. (2007). Modeling the leakage current of dielectric isolation structures in a high-voltage semiconductor technology. In *2007 IEEE International Symposium on Industrial Electronics* (pp. 1430–1434).
29. Lascar Electronics. PSU 30205 datasheet. Retrieved November 26, 2020, from https://www.lascarelectronics.com/media/7253/psu-30205.pdf
30. Le, H.-P., Sanders, S. R., & Alon, E. (2011). Design techniques for fully integrated switched-capacitor DC-DC converters. *IEEE Journal of Solid-State Circuits, 46*(9), 2120–2131.

31. Li, G., & Wu, X. (2022). A 98.4% efficiency 380 V–12 V DCX with 1.3 kW/in3 power density using low NFoM devices and resonant drive transformer. *IEEE Transactions on Power Electronics, 37*(10), 12346–12356.
32. Lu, Y., Jiang, J., & Ki, W.-H. (2017). A multiphase switched-capacitor DC–DC converter ring with fast transient response and small ripple. *IEEE Journal of Solid-State Circuits, 52*(2), 579–591.
33. Lutz, D. (2020). Integrated high-voltage switched-capacitor DC-DC converters. Ph.D. Dissertation, Leibniz University Hannover.
34. Lutz, D., Renz, P., & Wicht, B. (2016). 12.4 A 10 mW fully integrated 2-to-13 V-input buck-boost SC converter with 81.5% peak efficiency. In *2016 IEEE International Solid-State Circuits Conference (ISSCC)* (pp. 224–225).
35. Lutz, D., Renz, P., & Wicht, B. (2018). An integrated 3-mW 120/230-V AC mains micropower supply. *IEEE Journal of Emerging and Selected Topics in Power Electronics, 6*(2), 581–591.
36. Lutz, D., Seidel, A., & Wicht, B. (2018). A 50 V, 1.45 ns, 4.1 pJ high-speed low-power level shifter for high-voltage DCDC converters. In *ESSCIRC 2018 - IEEE 44th European Solid State Circuits Conference (ESSCIRC)* (pp. 126–129).
37. Makowski, M., & Maksimovic, D. Performance limits of switched-capacitor DC-DC converters. In *Proceedings of PESC '95 - Power Electronics Specialist Conference* (Vol. 2, pp. 1215–1221).
38. McPherson, J., Kim, J., Shanware, A., Mogul, H., & Rodriguez, J. (2002). Proposed universal relationship between dielectric breakdown and dielectric constant. In *Digest. International Electron Devices Meeting* (pp. 633–636).
39. McPherson, J., Kim, J., Shanware, A., Mogul, H., & Rodriguez, J. (2003). Trends in the ultimate breakdown strength of high dielectric-constant materials. *IEEE Transactions on Electron Devices 50*(8), 1771–1778.
40. Meyvaert, H., Sarafianos, A., Butzen, N., & Steyaert, M. (2014). Monolithic switched-capacitor DC-DC towards high voltage conversion ratios. In *2014 IEEE 15th Workshop on Control and Modeling for Power Electronics (COMPEL)* (pp. 1–5).
41. Meyvaert, H., Smeets, P., & Steyaert, M. (2013). A 265 V_{RMS} Mains Interface Integrated in 0.35 μm CMOS. *IEEE Journal of Solid-State Circuits, 48*(7), 1558–1564.
42. Meyvaert, H., Van Breussegem, T., & Steyaert, M. (2013). A 1.65 W fully integrated 90 nm bulk cmos capacitive DC-DC converter with intrinsic charge recycling. *IEEE Transactions on Power Electronics, 28*(9), 4327–4334.
43. Meyvaert, H., Villar Piqué, G., Karadi, R., Bergveld, H. J., & Steyaert, M. S. J. A light-load-efficient 11/1 switched-capacitor DC-DC converter with 94.7% efficiency while delivering 100 mW at 3.3 V. *IEEE Journal of Solid-State Circuits, 50*(12), 2849–2860.
44. Microchip. LR645. Retrieved July 12, 2022, from https://www.microchip.com/en-us/product/lr645
45. Mouser. Standard LEDs - SMD. Retrieved November 30, 2023, from https://eu.mouser.com/c/optoelectronics/led-lighting/led-emitters/standard-leds-smd/
46. Mouser. Vicor robitics application products. Retrieved December 29, 2023, from https://gr.mouser.com/new/vicor/vicor-robotics-application-products/
47. Murari, B., Bertotti, F., & Vignola, G. (2002). *Smart power ICs*. Springer.
48. OnSemi. OnSemi manufacturing processes datasheets. Retrieved August 26, 2021, from https://www.onsemi.com/products/product-taxonomy/custom-foundry-services/manufacturing-processes
49. Pasternak, S. R., Kiani, M. H., Rentmeister, J. S., & Stauth, J. T. Modeling and performance limits of switched-capacitor DC–DC converters capable of resonant operation with a single inductor. *IEEE Journal of Emerging and Selected Topics in Power Electronics, 5*(4), 1746–1760.
50. Paul Shepard. Ultra-high voltage foundry process for industrial and power applications. Retrieved November 13, 2023, from https://eepower.com/news/ultra-high-voltage-foundry-process-for-industrial-and-power-applications/

51. Pomper, M., Leipold, L., Muller, R., & Weidlich, R. (1978). On-chip power supply for 110 V line input. *IEEE Journal of Solid-State Circuits, 13*(6), 882–886.
52. Power Electronics News. The emerging 48 V ecosystem. Retrieved March 3, 2023, from https://www.powerelectronicsnews.com/the-emerging-48v-ecosystem
53. Pratt, A., Kumar, P., & Aldridge, T. V. (2007). Evaluation of 400 V DC distribution in telco and data centers to improve energy efficiency. In *INTELEC 07 - 29th International Telecommunications Energy Conference* (pp. 32–39).
54. Ramadass, Y., Blanco, A., Xiao, B., & Cummings, J. (2020). 18.3 A 120 mA non-isolated capacitor-drop AC/DC power supply. In *2020 IEEE International Solid- State Circuits Conference - (ISSCC)* (pp. 290–292).
55. Ramadass, Y. K., Fayed, A. A., & Chandrakasan, A. P. (2010). A fully-integrated switched-capacitor step-down DC-DC converter with digital capacitance modulation in 45 nm CMOS. *IEEE Journal of Solid-State Circuits, 45*(12), 2557–2565.
56. Razavi, B. (2015). The strongARM latch [a circuit for all seasons]. *IEEE Solid-State Circuits Magazine, 7*(2), 12–17.
57. RECOM. AC/DC converter RAC01-GA datasheet. Retrieved August 9, 2023, from https://docs.rs-online.com/d81a/0900766b815e16d9.pdf
58. Rindfleisch, C., Otten, J., & Wicht, B. (2022). A highly-integrated 20–300 V 0.5 W active-clamp flyback DCDC converter with 76.7% peak efficiency. In *2022 IEEE Custom Integrated Circuits Conference (CICC)* (pp. 1–2).
59. Rindfleisch, C., & Wicht, B. (2021). A Resonant one-step 325 V to 3.3–10 V DC–DC converter with integrated power stage benefiting from high-voltage loss-reduction techniques. *IEEE Journal of Solid-State Circuits, 56*(11), 3511–3520 (2021)
60. Rindfleisch, C., & Wicht, B. (2022). A 110/230 V AC and 15–400 V DC 0.3 W power-supply IC with integrated active zero-crossing buffer. *IEEE Journal of Solid-State Circuits, 57*(12), 3816–3824.
61. Rose, M., & Bergveld, H. J. (2016). Integration trends in monolithic power ICs: Application and technology challenges. *IEEE Journal of Solid-State Circuits, 51*(9), 1965–1974.
62. RS. A complete guide to LED drivers. Retrieved November 4, 2023, from https://uk.rs-online.com/web/content/discovery/ideas-and-advice/led-drivers-guide
63. Salem, L. G., & Mercier, P. P. (2014). A recursive switched-capacitor DC-DC converter achieving $2^N - 1$ ratios with high efficiency over a wide output voltage range. *IEEE Journal of Solid-State Circuits, 49*(12), 2773–2787.
64. Sarafianos, A., & Steyaert, M. (2015). Fully integrated wide input voltage range capacitive DC-DC converters: The folding dickson converter. *IEEE Journal of Solid-State Circuits, 50*(7), 1560–1570.
65. Schaef, C., Weng, S., Choi, B., Lambert, W., Radhakrishnan, K., Ravichandran, K., Tschanz, J., & De, V. (2019) 8.1 a 93.8% peak efficiency, 5 V-input, 10 A max ILOAD flying capacitor multilevel converter in 22 nm CMOS featuring wide output voltage range and flying capacitor precharging. In *2019 IEEE International Solid-State Circuits Conference - (ISSCC)* (pp. 146–148).
66. Seeman, M. D. (2009). A design methodology for switched-capacitor DC-DC converters. Ph.D. Dissertation, Department of Electrical Engineering and Computer Sciences (EECS), University of California.
67. Somasekhar, D., Srinivasan, B., Pandya, G., Hamzaoglu, F., Khellah, M., Karnik, T., & Zhang, K. (2009). Multi-phase 1 GHz voltage doubler charge-pump in 32 nm logic process. In *2009 Symposium on VLSI Circuits* (pp. 196–197).
68. Stauth, J. T. (2018). Pathways to mm-scale DC-DC converters: Trends, opportunities, and limitations. In *2018 IEEE Custom Integrated Circuits Conference (CICC)* (pp. 1–8).
69. Steyaert, M., et al. (2022). DCDC performance survey. Retrieved July 12, 2022, from https://bit.ly/3S27GC5
70. STMicroelectronics. BCD technologies. Retrieved November 13, 2023, from https://www.st.com/content/st_com/en/about/innovation---technology/BCD.html

71. Su, L., Ma, D., & Brokaw, A. P. (2008). A monolithic step-down SC power converter with frequency-programmable subthreshold z-domain DPWM control for ultra-low power microsystems. In *ESSCIRC 2008 - 34th European Solid-State Circuits Conference* (pp. 58–61).
72. Sullivan, C. R., Reese, B. A., Stein, A. L. F., & Kyaw, P. A. (2016). On size and magnetics: Why small efficient power inductors are rare. In *2016 International Symposium on 3D Power Electronics Integration and Manufacturing (3D-PEIM)* (pp. 1–23).
73. Tamez, A. A., Fredenburg, J. A., & Flynn, M. P. (2010). An integrated 120 volt AC mains voltage interface in standard 130 nm CMOS. In *2010 Proceedings of ESSCIRC* (pp. 238–241).
74. Tan, Y. K., & Panda, S. K. Energy harvesting from hybrid indoor ambient light and thermal energy sources for enhanced performance of wireless sensor nodes. *IEEE Transactions on Industrial Electronics, 58*(9), 4424–4435.
75. Tanzawa, T., & Tanaka, T. (1997). A dynamic analysis of the Dickson charge pump circuit. *IEEE Journal of Solid-State Circuits, 32*(8), 1231–1240.
76. Texas Instruments. Design guide: TIDA-010060. Retrieved July 12, 2022, from https://www.ti.com/lit/ug/tidueq2/tidueq2.pdf
77. Thielemans, T. (2023). Fully integrated switched capacitor converters towards high-density capacitive energy storage. Ph.D. Dissertation, ESAT-MICAS, KU Leuven.
78. Thielemans, T., Butzen, N., Sarafianos, A., Steyaert, M., & Tavernier, F. (2018). A capacitive DC-DC converter for stacked loads with wide range DVS achieving 98.2% peak efficiency in 40 nm CMOS. In *2018 IEEE Custom Integrated Circuits Conference (CICC)* (pp. 1–4).
79. Tong, T., Lee, S. K., Zhang, X., Brooks, D., & Wei, G.-Y. A fully integrated reconfigurable switched-capacitor DC-DC converter with four stacked output channels for voltage stacking applications. *IEEE Journal of Solid-State Circuits, 51*(9), 2142–2152.
80. Van Breussegem, T., & Steyaert, M. (2010). A fully integrated gearbox capacitive DC/DC-converter in 90 nm CMOS: Optimization, control and measurements. In *2010 IEEE 12th Workshop on Control and Modeling for Power Electronics (COMPEL)* (pp. 1–5).
81. Van Breussegem, T., & Steyaert, M. (2013). *CMOS integrated capacitive DC-DC converters*. Springer.
82. Van Breussegem, T. M., & Steyaert, M. S. J. (2011). Monolithic capacitive DC-DC converter with single boundary–multiphase control and voltage domain stacking in 90 nm CMOS. *IEEE Journal of Solid-State Circuits, 46*(7), 1715–1727.
83. Van Daele, T., & Tavernier, F. (2022). A 400-to-12 V fully integrated switched-capacitor DC-DC converter achieving 119 mW/mm^2 at 63.6 % efficiency. In *2022 IEEE Custom Integrated Circuits Conference (CICC)* (pp. 1–2).
84. Van Daele, T., & Tavernier, F. (2023). A fully integrated 230 VRMS-to-12 VDC AC-DC converter achieving 9 mW/mm^2. In *2023 IEEE Symposium on VLSI Technology and Circuits (VLSI Technology and Circuits)* (pp. 1–2).
85. Van Daele, T., & Tavernier, F. (2023). Fully integrating a 400 V-to-12 V DC–DC converter in high-voltage CMOS. *IEEE Journal of Solid-State Circuits, 58*(3), 732–741.
86. Van Daele, T., & Tavernier, F. (2023). Monolithic 230-VRMS-to-12-VDC AC–DC converter at 9 mW/mm^2 enabled by a 31–325-VDC input range capacitive multi-ratio DC–DC converter. *IEEE Journal of Solid-State Circuits*.
87. Van Daele, T., & Tavernier, F. (2023). Overcoming the challenge of fully integrated DC-DC converters with high input voltages in CMOS. In: K. Makinwa, A. Baschirotto & B. Nauta (Eds.), *Imaging Sensors, Power Management, PLLs, and Frequency Synthesizers - Advances in Analog Circuit Design 2023*. Springer
88. Veraverbeke, B., Thielemans, T., Daele, T. V., & Tavernier, F. (2022). A 240 V to 47.5 V fully integrated switched-capacitor converter in GaN achieving 62.6% efficiency at 220 mW/mm^2. In *2022 17th Conference on Ph.D Research in Microelectronics and Electronics (PRIME)* (pp. 249–252).
89. Wei, K., Ramadass, Y., & Ma, D. B. (2020). 11.1 a direct 12 V/24 V-to-1 V 3 W 91.2%-efficiency tri-state DSD power converter with online VCF rebalancing and in-situ precharge

rate regulation. In *2020 IEEE International Solid- State Circuits Conference - (ISSCC)* (pp. 190–192).
90. Wittmann, J. (2020). *Integrated high-Vin multi-MHz converters*. Springer.
91. Wittmann, J., Funk, T., Rosahl, T., & Wicht, B. (2018). A 48-V wide- V_{in} 9–25-MHz resonant DC–DC converter. *IEEE Journal of Solid-State Circuits, 53*(7), 1936–1944.
92. World Standards. Plug, socket and voltage by country. Retrieved March 2, 2023, from https://www.worldstandards.eu/electricity/plug-voltage-by-country/
93. X-FAB. X-FAB high-voltage technologies. Retrieved November 13, 2023, from https://www.xfab.com/technology/high-voltage
94. Xue, J., & Lee, H. (2016). A 2 MHz 12–100 V 90% efficiency self-balancing ZVS reconfigurable three-level DC-DC regulator with constant-frequency adaptive-on-time V^2 control and nanosecond-scale ZVS turn-on delay. *IEEE Journal of Solid-State Circuits, 51*(12), 2854–2866.
95. Yamauchi, Y., Sai, T., Hata, K., & Takamiya, M. (2021). 0.55 W, 88%, 78 kHz, 48 V-to-5 V fibonacci hybrid DC–DC converter IC Using 66 mm^3 of passive components with automatic change of converter topology and duty ratio for cold-crank transient. *IEEE Transactions on Power Electronics, 36*(8), 9273–9284.

Index

© The Editor(s) (if applicable) and The Author(s), under exclusive license to Springer Nature Switzerland AG 2025

T. Van Daele, F. Tavernier, *Fully Integrated High-Voltage DC-DC and AC-DC Conversion*, Analog Circuits and Signal Processing,
https://doi.org/10.1007/978-3-031-78519-1

The manufacturer's authorised representative in the EU is Springer Nature Customer Service Centre GmbH, Europaplatz 3, 69115 Heidelberg, Germany. If you have any concerns regarding our products, please contact ProductSafety@springernature.com

Printed and bound by CPI Group (UK) Ltd, Croydon, CR0 4YY
30/09/2026
02225607-0007